# Environmental and Health Regulation in the United States and the European Union

# Environmental and Health Regulation in the United States and the European Union

# Protecting Public and Planet

Mitchell P. Smith

palgrave
macmillan

ENVIRONMENTAL AND HEALTH REGULATION IN THE UNITED STATES AND THE EUROPEAN UNION


First published in 2012 by
PALGRAVE MACMILLAN®
in the United States—a division of St. Martin's Press LLC,
175 Fifth Avenue, New York, NY 10010.

Where this book is distributed in the UK, Europe and the rest of the world, this is by Palgrave Macmillan, a division of Macmillan Publishers Limited, registered in England, company number 785998, of Houndmills, Basingstoke, Hampshire RG21 6XS.

Palgrave Macmillan is the global academic imprint of the above companies and has companies and representatives throughout the world.



ISBN: 978–0–230–10936–0

Library of Congress Cataloging-in-Publication Data

Smith, Mitchell P., 1960–
Environmental and health regulation in the United States and the European Union : protecting public and planet / Mitchell P. Smith.
p. cm.
Includes bibliographical references.
ISBN 978–0–230–10936–0 (hardback)
1. Environmental policy—United States. 2. Environmental policy—European Union countries. 3. Medical policy—United States. 4. Medical policy—European Union countries. I. Title.

GE180.S44 2012
363.7'056094—dc23 2011034898

A catalogue record of the book is available from the British Library.

Design by Newgen Imaging Systems (P) Ltd., Chennai, India.

First edition: March 2012

# Contents

# Figures and Tables

## Figures



## Tables

# Acknowledgments

It is always a bit startling to reflect at the end of a book project on the number of years that have passed from initiation to completion. I have incurred immense debts of gratitude throughout these past five years, during which I've wended my way through several distinct areas of research in both the US and European contexts. At the same time, the five years have been an enjoyable experience not only due to the rewards of the work, but also because of the generosity and professionalism of numerous practitioners, scholars, colleagues, and students.

I began the project in 2006 immersed in the fascinating but intricate debate over the EU's proposed chemicals regulation, REACH. Many people helped me move the project forward, whether through sharing their expertise, offering thoughtful feedback on early versions of my work, or providing financial support. Expertise came in particular from representatives of environmental interest associations, chemical industry employers, and EU institutions, all of whom were generous with their time and knowledge during a spring 2006 research visit to Brussels at the project's inception. Mecki Naschke, then of the European Environmental Bureau's EU Policy Unit; Aleksandra Kordecka of Friends of the Earth Europe; and Delia Villagrasa, at the time European Parliamentary Affairs officer for World Wide Fund for Nature Europe, all provided critical and impressively informed insight from the perspective of environmental NGOs. Franco Bisegna, Senior Counsellor for EU Government Affairs at CEFIC, the European Chemical Industry Council, gave expansive and illuminating answers to questions about

REACH both by e-mail and in person. Within the European Parliament, Sabina Magnano offered enlightenment on the context of the REACH debate. Jochen Richter provided tremendous insight into the intricacies of parliamentary procedures involved in the treatment of the REACH dossier in committees, as did the eminently knowledgeable Francis Jacobs.

Many scholarly colleagues and friends on both sides of the Atlantic have fostered the intellectually fertile environment in which I've been fortunate to operate. My sincere thanks to David Howarth, who organized and hosted an April 2006 workshop on Economic Interests and European Integration at the University of Edinburgh that allowed me to present a preliminary version of my research on REACH. Along with Tal Sadeh, co-organizer of the endeavor, David was an exemplary research project leader. For their quality scholarship and collegiality, I extend my gratitude to fellow workshop participants David Coen, Andreas Dür, David Ellison, Emiliano Grossman, Alexandra Hennessy, Patrick Leblond, Lucia Quaglia, Amy Verdun, Ramūnas Vilpišauskas, and Cornelia Woll. Tony Zito of the University of Newcastle provided constructive critique at the Edinburgh workshop, and thoughtfully commented on subsequent versions of the work on REACH, lending insight from his own highly accomplished work on EU environmental policy. Charlotte (Charlie) Burns of the University of Leeds offered substantive feedback on an early paper on REACH that formed the core of the EU portion of chapter 2. Joel Tickner of the School of Health & Environment at the University of Massachusetts Lowell also provided insightful comments on an early REACH draft.

To my delight, David Howarth invited me back to the University of Edinburgh in his capacity as Jean Monnet Chair to present my work in progress on the EU's regulatory trilemma and was again a gracious host. Thanks to Mark Aspinwall and Chad Damro, who contributed productive critique of that very preliminary work and helped make the spring 2009 Edinburgh visit truly memorable.

At the University of Oklahoma, Kyle Erwin provided research assistance with information on regulation of health claims on food labels in the United States. I express my gratitude to the Carl Albert Center, which funded Kyle's Undergraduate Research Fellowship. Others at the University of Oklahoma have granted consistent and generous funding for my travel to conduct research or present portions of the work at various venues. The Office of the Vice President for Research, the Office of the Dean of the College of Arts and Sciences, and the College of International Studies all lent repeated support to my work, permitting me to participate in the spring 2006 Edinburgh workshop on Economic Interests and the Construction of Europe, to present an element of the research at the Third Pan-European Consortium for Political Research in Istanbul in fall 2006 and to deliver another paper from the project at the International Studies Association (ISA) meeting in San Francisco in March 2008. Underscoring this dedicated support for research at the University of Oklahoma, all three offices provided travel funding for me to present my work at the April 2009 Eleventh Biennial Conference of the European Union Studies Association in Los Angeles.

Serving as discussant on a panel on international economic agreements at the 2008 ISA conference, Alan Cafruny provided valuable comments on my regulatory trilemma argument. Fellow panelists Herbert Zimmerman and Ljiliana Biukovic also offered extremely useful feedback.

I have enjoyed numerous discussions at conferences in both the United States and Europe with Steven Kennedy of Palgrave Macmillan; I admire his vision as a publishing director and his dogged encouragement of scholarly productivity and thank him for steering my work toward Robyn Curtis at Palgrave. My thanks also to Robyn for her patience while my effort to bring the book to completion stretched out over several months.

Tom Banchoff and Tom Glenn have been cherished friends and sources of stalwart support throughout the course of this project and long before as well. Thanks to Gretchen Samuels, world's best and most dedicated kickboxing instructor, for keeping me fit despite countless hours sitting with the laptop.

My deepest gratitude is to my family, to whom this book is dedicated. I am so thankful for the constant loving support of my mom, Cynthia Rosen. My wonderful parents-in-law, Penny and Brian Williams, have given their love and encouragement for many years. Thanks with love to Micah and Megan; your humor is a constant joy and your creative talents an endless source of pride. Most of all, I could not have accomplished anything without the love, support, and intellectual and life companionship of my wife, Lynn C. Lewis. And, finally, thanks to little Willow, such loyal and comforting canine company stretched out under the hassock as I worked through final revisions in my favorite chair. You deserved every treat.

CHAPTER 1

# DIVERGENT REGULATORY TRAJECTORIES

DURING THE FIRST DECADE OF THE TWENTY-FIRST CENTURY, the United States adopted an increasingly relaxed regulatory posture in the face of critical challenges to public health and the environment. This is true of regulation of reuse and recycling of end-of-life products including autos and electronic components, of potentially hazardous chemicals, and of health claims on food labels. Coincidentally, the European Union (EU) gravitated toward more restrictive regulation in these very same areas, establishing more stringent controls on the recycling of autos at the end of their useful lives (2000); tighter rules for health claims on food labels (2005); and a more rigorous regime for regulation of chemicals (2006). How might we explain these diverging regulatory trajectories of the world's two largest market economies in an era of rising public awareness of dangers to the public and the planet?

Observers have taken note of the contrasting directions of health and environmental regulation in the United States and the EU,[1] offering several explanations. The most prominent of these refers to systematic differences in societal willingness to tolerate risk, or in cultural perceptions of types of risk considered tolerable. Such arguments range from those emphasizing European responses to food scares in the recent past, including the instance of bovine spongiform encephalopathy (BSE) in beef and dioxin in carbonated beverages, poultry, and eggs in 1999, to those suggesting Europeans are accepting of traditional foods but more suspicious of new technologies than American consumers—exhibited, for example, in differences in American and European attitudes toward genetically modified foods (Echols, 1998; Dunlop, 2000; Rosendal, 2005; Kurzer, 2005). But there is a fundamental problem with such accounts: cultural explanations resting on systematic attitudinal differences across populations fail to account for the low level of risk tolerance and high level of precaution exhibited by past regulatory policy in the United States—including, for example, a virtual ban on health claims on food labels for most of the twentieth century (Pappas, 2002)—as well as more recent policy in a variety of areas, including tobacco consumption (Kurzer, 2005), blood donations from individuals potentially exposed to BSE beef (Wiener and Rogers, 2002), and regulation of nitrogen oxides from diesel vehicles (Oye, 2005: 62). These cultural explanations also neglect the frequency with which regulatory regimes in European countries in a variety of areas involving potentially hazardous and toxic materials resembled US regulation—with reliance on voluntary partnerships between public authorities, environmental interest associations, and industry—prior to regulation at the EU level.

An emphasis on characteristic features of political economies, particularly the balance between state and market, is an alternative to the cultural explanation. This approach underscores the distinction between business-centered, liberal US capitalism and state-centered European dirigisme. Deeply institutionalized relationships between state and market mean that the political economy of the United States systematically produces regulatory structures that give primacy to free competition; in contrast, Europe's coordinated capitalism involves a significantly greater willingness to constrain market exchange in order to generate public goods (Krämer, 2004: 68).

This explanation of diverging regulatory trajectories confronts the fact that the construction of an integrated European market has strongly favored intensification of market competition. Policy debates both at national level and within EU institutions exhibit a deep and widely shared concern with the cost burden of regulation on industry. Reflecting this dimension of the legislative process in the EU, regulatory impact assessment has come to play an increasingly prominent role in the drafting of regulation (Radaelli, 2007; Cecot et al., 2008). In this regard, EU discourse and policy debate closely resemble that in the United States.

Advocates of the state-market balance explanation also refer to the role of courts in bolstering market competition in the US setting. As I explain below, court decisions that have fleshed out the practical meaning of regulations in the United States have indeed tended to impose extremely high evidentiary standards on regulators like the Food and Drug Administration (FDA) to justify their rulings and to demand that authorities restrain regulatory solutions to the least intrusive means required to achieve the health or environmental objectives they seek to attain. This has made outright bans by the FDA on dangerous substances (asbestos, for example) or on claims (as on food labels) difficult to sustain. In addition, courts have produced an expansive interpretation of free commercial speech doctrine—limiting, for example, the ability of regulators to constrain product claims. The striking point, though, is that *decisions of the European Court of Justice (ECJ) follow this pattern remarkably closely*. Proportionality and free commercial speech doctrine—justified by the EU's constitutionally established and essential principle of free movement of goods—have become important constraints on national regulation in EU member states. In other words, preliminary rulings by the ECJ repeatedly have rejected interpretations of national regulations that restrict the free movement of goods and services across borders. Ultimately, contrasting regulatory trends between the United States and the EU cannot be ascribed to an institutional proclivity in Europe to constrain market exchange in favor of the public welfare.

Finally, scholars have emphasized the role of the industry lobby in the United States—both its organizational resources and its access to regulatory policy making (Tanguay, Lanoie, and Moreau, 2004). In contrast, some scholars studying the dynamics of policy making and representation in the EU have emphasized the significant access of organized environmental interests to policy making in the EU, whether through the European Parliament's (EP) Committee on the Environment (Collins, Burns, and Warleigh, 1998; Keading, 2004), the Environment Directorate General of the European Commission, or environment ministries in high-standard EU member state governments. However, two crucial realities confront this explanation for contrasting trajectories. First, whatever the impact of industry lobbying on regulatory policy making in the United States, court decisions have been the decisive factor in determining the consequences of regulatory design in the areas of chemicals and nutrition labeling regulation. The history of nutrition labeling regulation dating back to the start of the twentieth century, for example, demonstrates a deep embedding of the principle of precaution and a sustained adherence to this principle by the US FDA despite a succession of legislative measures, which are designed to relax the regulatory regime and which reflect industry's access to Congress.

A second development undermining the pluralist industry lobbying explanation for diminished rigor of health and environmental regulation in the United States compared with that in the European Union is that, in the case of the EU, access to the policy-making process of industry interests organized at the European level has expanded substantially as the powers of the EP have grown (Bouwen, 2004; Eising, 2007: 350, 352). As the EP has developed full co-decision authority on legislation, industry associations have invested more substantial organizational resources in and have identified productive access points within the Parliament (Coen, 2007).[2] For example, the close examination in chapter 2 of the process that generated REACH, the EU's recent comprehensive regulatory framework for the chemicals sector, clearly reveals a shift over time from the ability of environmental interests to set the agenda for environmental regulation to augmented industry influence over the positions of the EP. The account of the development of regulation to restrict health claims on food labels in chapter 3 shows that food producers had substantial success gaining support in the EP for their efforts to block a central provision of the legislation that would severely restrict claims—a requirement that products for which claims could be made not exceed certain levels of salt, sugar, and fat—even though the EP's position ultimately did not prevail. And in regulation of recycling of vehicles reaching the end of their useful lives, as discussed in chapter 4, auto industry lobbying complicated decision making in the Council of Environment Ministers, shaped amendments to the Council's position proposed and passed by the EP, and altered the course of implementation of EU regulation.

Regulatory distinctions between the EU and the United States are not simply a matter of degree; there also are critical differences of kind. In particular, voluntary agreements between industry and government proliferate in the United States, whereas the European context has tended to move from voluntary regulatory arrangements at the national level—as in the case of arrangements in the United Kingdom and Sweden for health claims on food labels, or the German system for recycling end-of-life vehicles (ELVs)—to uniform and compulsory EU-level regulation. This book explains the contrasting regulatory trajectories of the United States and the EU in terms of two factors. First, all capitalist democracies confront a perceived trade-off between regulatory objectives and costs to industry. Particularly in traded sectors, such as chemicals, policy makers face a "regulator's dilemma" emerging from the tension between regulatory objectives and international competitiveness

(Kapstein, 1989). However, in the case of the EU, the trade-off between regulatory objectives and costs is a regulatory *trilemma*. The standard trade-off is compounded for the EU by the fact that regulation also serves a goal beyond immediate regulatory objectives such as protection of environment, health, or consumers—the goal of advancing European integration. Understanding the regulatory trilemma helps us comprehend critical features of regulation in the EU, including a tendency (though by no means unrestrained or unyielding) toward mandatory regulation rather than voluntary regulation.[3]

In addition to the contrast between a classic regulator's dilemma in the United States and the EU's regulatory trilemma, the implications of court decisions represent a second explanatory factor for diverging United States and EU regulatory trajectories. Courts have sharply constrained federal regulatory ambitions in the United States. This is particularly clear from the evidentiary standards imposed by courts on the FDA to provide cost-benefit analyses of its regulatory remedies in the chemicals sector. The constraining role of courts also is evident even where compulsory regulation prevails in the United States; decisions protecting free commercial speech associated with health claims on food labels have severely limited the degree of consumer protection afforded by regulation.

In contrast, rulings issued by the ECJ, while coinciding with those in the United States in terms of their tendency to protect open markets and free commercial speech, have had a very different function. Court rulings typically have rejected highly restrictive national regulations ostensibly designed to protect consumers on the grounds that they constrain the free flow of goods across borders. This was true, for example, of the 1987 decision on the compatibility of German beer purity laws with Europe's single market.[4] Similarly, the ECJ in 2002 and 2003 ruled respectively against a 1975 Austrian law on nutrition labeling that imposed an outright ban on health claims, and a similar Belgian law of 1980, indicating that in both cases the law's impact was disproportionate to its stated consumer-protection objective.[5] These decisions have constrained national consumer-protection measures as they have in the United States. However, *within the EU, rulings undermining national regulatory rigor have served as catalysts for EU institutions to seek to expand the EU's regulatory ambit by sharply delineating areas where single market regulation remains incomplete.* In the case of nutrition labeling, for example, the European Commission in July 2003 put forward a proposal for a new health claims regulation that would permit health claims only under highly restrictive conditions that would be uniform across EU member states.[6] The proposal drew substantial (if not uniform) support from national health ministers comprising the EU Council of Health Ministers, who endorsed European rules limiting health claims that adhered closely to some of the very national measures struck down by decisions of the ECJ.

In the following section, I elaborate on the first factor explaining the divergent regulatory trajectories of the United States and the EU in the areas of health and the environment—the EU's regulatory trilemma. I then assess the contrasting impact of court decisions in the two settings. I also highlight the growth over time of industry access to critical decision-making nodes in the EU policy process and the rise of regulatory impact assessment as an indicator of rising influence of organized industry interests. I conclude this introductory chapter by setting out the organization of the book and identifying the focus of each subsequent chapter.

## The EU's Regulatory Trilemma

All states with open economies run the risk of creating externalities for domestic firms when they regulate markets. Ethan Kapstein refers to this tension between regulatory objectives and international competitiveness in traded sectors as the "regulators' dilemma" (Kapstein, 1989). International political economy scholars have tended to focus on efforts to resolve the dilemma through uploading of domestic regulatory regimes through international agreement. Robert Falkner, for example, examines the effort of the EU in the late 1990s to export its domestic policy on regulation of genetically modified organisms, suggesting this process was "motivated by a desire to secure international legitimacy for the EU's own precautionary approach" (Falkner, 2007: 520). In his study of EU environmental policy in a competitive global economy, Jonathan Golub argues along related lines: "Negotiating international environmental agreements allows EU members to level the economic playing field and to undermine the effects of pollution havens" (Golub, 1998: 5).

For the EU, the classic trade-off between regulatory objectives and economic competitiveness is in fact a regulatory trilemma, in which the European integration objective compounds the trade-off between regulatory goals and costs to industry. This is depicted in figure 1.1 below.

EU regulation is a tool of integration because it induces interest articulation at the European level, harmonizes standards, and intensifies exchange across borders of EU member states. As EU institutions legislate, they stimulate and sustain interest organization and articulation at the European level (Marks and McAdam, 1996; Kohler-Koch, 1997; Pollack, 1997; Fairbrass and Jordan, 2001; Mahoney, 2004). This is true of trans-European interest associations organized on behalf of the

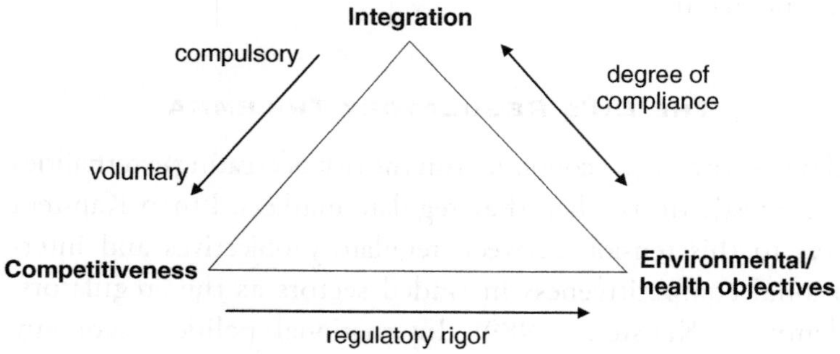


Figure 1.1 The European Union's regulatory trilemma.

environment, public health, or consumers, as well as associations of business organized by industrial sectors or according to the challenges they face in meeting regulatory burdens (small and medium enterprises, for example). Debates over regulatory outcomes at the EU level themselves serve the objective of integration. European-level interest aggregation and articulation and policy debate constitute what we might term the mobilization benefits of EU regulatory policy making.

The EU's regulatory trilemma produces characteristics of regulatory behavior that differ in critical respects from regulation at the national level in economic competitors of the EU, especially the United States. For example, the US regulatory structure may accommodate industry resistance to command and control regulation through arrangements for voluntary regulation, including programs that rely on the impact of information and moral suasion, such as the toxic release inventory (TRI). This also may be accomplished through regulatory preemption by industry, typically consisting of cross-industry consortiums of firms cooperating with government agencies to encourage improved environmental stewardship and intensified research and development focused on cost-effective forms of environmental remediation.

There is a debate in the regulation literature about the effectiveness of voluntary regulations in meeting regulatory objectives[7]; whatever the impact on regulatory outcomes,[8] voluntary agreements may not always be a productive means of advancing European integration. Consequently, EU institutions are less inclined than US regulators to rely on voluntary regulation and more inclined toward compulsory regulation.[9] Among other examples, a comparison of recent measures to regulate potentially dangerous chemicals in the United States (a combination of industry-led schemes for voluntary commitment to good product stewardship and incremental regulations in a small number of states) and the EU (a compulsory, EU-wide system for registration, evaluation, and authorization of chemicals) supports the argument.[10]

It is critical to note that the integration-driven tendency toward compulsory regulation does not mean that implementation and enforcement proceed seamlessly in the EU context. In fact, substantial noncompliance is a critical functional component of augmented regulatory rigor in the EU. Indeed, the EU's regulatory trilemma may be resolved in any of three ways: regulatory accommodation; compliance deficits; or uploading of EU rules to the international level (Smith, 2010). Regulatory accommodation occurs when initial aspirations for environmental or health protection are relaxed in the regulatory policy-making process in response to demands from institutional actors or organized interests. Compliance deficits occur when EU institutions produce rigorous regulation—such as when national environment or health ministers pursue an opportunity to recapture regulatory control at the European level—but compliance costs are initially high. While observers implicitly assume that noncompliance reflects a shortcoming of EU regulatory capacity, compliance deficits may be a functional outcome of regulatory policy making and, indeed, an equilibrium state for a period of time provided noncompliance does not directly jeopardize progress toward market integration. Put differently, as reflected in figure 1.1, European-level regulation may under some conditions advance integration independently of the level of compliance.

Finally, given its status as a market leader in some sectors, there may be no need for the EU to upload regulatory regimes in order to set global standards. This is the case in the chemicals sector, for example. In other sectors, where the EU does not wield sufficient market clout to impose on global economic actors the requirement of compliance with its regulatory regime, EU institutional actors may seek to transmit EU rules to the global level. This is especially likely where a high level of compliance with the regulatory regime is essential to the effective functioning of Europe's single market. As an example, the European Commission has long sought the uploading of its open regime for public procurement to the international level through the General Agreement on Trade in Services component of the World Trade Organization. Having made little progress in this forum, the EU has turned more recently to pursuing this objective through bilateral trade agreements, as with Chile and South Korea.[11]

## Courts in the United States and the EU: Free Commercial Speech versus Single-Market Completion

The history of regulation of health claims on food labels in the United States illuminates the central role played by the precautionary principle, and also demonstrates that courts more than industry lobbying account for the recent decline in regulatory rigor. Many legal scholars portray the FDA as adhering rigidly to a stubbornly restrictive and paternalistic regulatory approach to health claims on food labels that dates back to the 1906 Federal Food and Drugs Act. Although over time Congress came to mediate the conflicting demands of consumer-protection advocates and food industry associations by enacting somewhat more permissive rules, the FDA adhered to its aversion to allowing broad use of health claims on food labels. The juxtaposition of a Congress that is at least somewhat responsive to industry interests and a regulatory agency that has persistently adhered to a strict interpretation of rules belies the notion of a culture of high risk tolerance.

Court decisions—rather than acts of Congress that may well reflect the influence of industry lobbies—have eroded restraints on nutrition labeling on free commercial speech grounds. As discussed in chapter 3, the landmark was the 1999 *Pearson v. Shalala* case, which challenged the FDA's application of the "significant scientific agreement" standard for approving health claims established in the 1993 Nutrition Labeling and Education Act (NLEA).[12] The decision was based on the court's interpretation of First Amendment protection of free commercial speech and the high threshold implied for suppression of such speech. The FDA nonetheless attempted to adhere to a strict interpretation of the provisions of the NLEA, but the case cracked open the door to a series of court decisions that progressively widened the opening to claims on food labels based on the notion that the FDA could find less restrictive ways to regulate claims—principally disclaimers rather than outright bans.

In the case of the EU, courts reached similar decisions to those in the United States on issues of proportionality of regulatory remedies and free commercial speech. However, crucial for the EU was that different standards across member states raised concerns in the courts about interference with the free flow of goods across borders (European Commission, 2001). In instances where governments enacted restrictive regulations governing health claims, they risked incompatibility with community law by interfering with the free movement of goods. A series of ECJ preliminary rulings in the early 2000s established that national rules providing for outright restrictions of health claims failed the test of proportionality. In *Commission v. Austria*, the court ruled that the total prohibition on health claims on food labels constituted by Austria's 1975 federal law on trade in foodstuffs violated the principle of proportionality with the objectives sought.[13] In its Douwe Egberts ruling, the court found that a 1980 Belgian law that had the effect of banning the weight control claim of a coffee product infringed EU law. As an alternative to outright prohibition of the claim, regulatory authorities had the option of employing the less restrictive approach of obligating the producer to supply evidence supporting the claim. The regulatory remedy was accordingly disproportionate to the objective of protecting human health.[14]

While the immediate effect of these cases was to constrain highly restrictive national regulations in order to protect open markets, the ultimate impact was to generate broad support from national governments as well as industry for regulation at the EU level. In its 2001 "Discussion Paper on Nutrition and Functional Claims," the European Commission's Health and Consumer Protection Directorate General emphasized that regulatory differences across member states could impede the free movement of food products and thereby inhibit the functioning of the internal market (European Commission, 2001: paragraph 4, p. 3). National health ministers meeting in the Council, meanwhile, sought to reestablish the regulatory controls eroded by preliminary rulings of the ECJ. In response to the preferences of national health ministers, the European Commission in 2003 proposed a regulation that would restrict the use of health claims on food packaging.

Industry representatives sought the legal certainty and facilitation of free movement of goods that would result from EU-level regulation, even if they pushed for less restrictive regulation than that advocated by some national health ministers. Demonstrating the rising influence of organized industry interests in the European Parliament, the EP stripped the Commission regulatory proposal of its most controversial element—a provision for nutrition profiles that would preclude health claims in cases of foods with high levels of fat, sugar, or salt. However, highlighting the power of ECJ rulings to mobilize national actors at the European level, the Council of Health Ministers succeeded in reintroducing nutrition profiles into the final regulation of December 2006.

This argument is consistent with literature on the role of the ECJ in expanding European-level environmental regulation. Critical here is the role of private litigants.[15] As this literature stresses, "[L]itigants disproportionately target national environmental laws which obstruct transnational activity" (Cichowski, 1998: 402). The consequence is that decisions at the European level dismantle national level regulatory regimes, shifting the foundations for establishment of rules upward to the supranational level (Cichowski, 1998: 402).

The first such instance concerns a late 1970s French government decree implementing a 1975 Council of Ministers

Directive on the collective and disposal of waste oils. The decree required that French firms dispose of waste oils at a regionally approved group of firms authorized to operate and accept such oils in the French market; the shipment abroad of waste oils to disposal centers authorized by other governments of EU single-market countries was prohibited. The European Court's preliminary ruling on the question forwarded from a French regional court established that environmental protection measures had to be narrowly tailored to their economic objectives in order to avoid clashing with the principle of free movement of goods.[16] By creating a monopoly position, the law would give French waste oil collectors a competitive advantage (Vedder, 2003: 270). As Rachel Cichowski points out, "This case typifies a series of waste rulings in which the Court systematically dismantles national environmental regulations which create an obstruction to transnational exchange" (Cichowski, 1998: 399). Such cases generate mobilization benefits as private litigants seek redress from the ECJ in pursuit of their economic interests: "These societal actors may not be actively pursuing European integration per se, but . . . the unintended consequences of their actions have a direct impact on the construction of supranational policy and the deepening of integration" (Cichowski, 1998: 403).

As with waste oils, the response of private competitors of food producers gaining economic advantage from national market rules ultimately shifted regulation of health claims on food labels upward to the EU level.

## Growing European Industry Access and Regulatory Impact Assessment

Early assessments of the impact of organized interests on EU policy making suggested the EU might provide a fertile environment for diffuse environmental interests, consumer-protection advocates, and women's rights groups. As is widely recognized, the access of business interests to the policy-making apparatus at the European level has increased substantially over time, involves multiple effective channels, and has had a traceable impact on the course of recent legislation. It is therefore not at all convincing to attribute the difference in EU and US regulatory trajectories to the relative weakness of the industry lobby in the EU.

Augmented industry influence is the product of a protracted process rather than a recent shift in policy making and interest representation in the EU. As David Coen points out, the shift in business lobbying resources toward the European level followed the 1986 Single European Act (SEA). Since the SEA placed decision making involving single-market issues on a weighted majority basis, industry was now confronted with the prospect of having their governments outvoted in the Council of Ministers; accordingly, the business sector could not rely exclusively on national channels of influence and had to diversify lobbying resources (Coen, 1997: 95). At the same time, the European Commission, faced with a broader and greater regulatory policy load, needed reliable providers of information on specific industrial sectors; large firms became leading actors in sectoral forums that provided the Commission with information resources and enhanced legitimacy, and also furnished it with a link to domestic policy-making environments (Coen, 1997: 96).

As Coen finds in his survey of large European firms, in this early post-SEA environment, industry recognized the growing legislative importance of the European Parliament, but was not yet prepared to devote additional scarce lobbying resources to the EP. As he asserts, "[W]hile the European Parliament attempts to establish a constituency with the people of Europe, the European Commission is contenting itself with developing a business constituency" (Coen, 1997: 104). However, within a few years, the growing legislative stature of the EP in fact induced the industry lobby to invest more heavily in efforts to influence EP committees.

Institutionally, two channels have facilitated augmented industry influence in EU policy making. The first of these is the formation of the Competitiveness Council, created at the 2002 Seville summit of heads of state and government through a merger of the Internal Market, Industry, and Research Councils. The Competitiveness Council, comprised of national ministers for economics, industry, and research, is charged with serving as a guardian of the competitiveness of European industry and ensuring that the Commission consults and takes account of the concerns of the business community in the policy-making process.[17] Member state governments identified the creation of the Competitiveness Council as "a response to the need for a more coherent and better co-ordinated handling of matters closely related to the competitiveness of European enterprises."[18]

The second institutional facilitator of enhanced industry influence is the EP's Internal Market Committee (IMCO), created at the start of the EP's sixth parliamentary term in July 2004. IMCO was given responsibility for coordination of legislation involving the internal market and (somewhat contradictorily) protecting the economic interests of consumers.[19]

Procedurally—and certainly not unrelated to the institutional innovation that has occurred—the extended use of impact assessment in EU policy making reflects an altered EU policy discourse in which: (1) the core objective is minimizing the regulatory cost burden imposed on industry; and (2) affected industry is central to the process of assessing the cost burden involved.[20] The Competitiveness Council has consistently advocated for more intensive reliance on regulatory impact assessment. In 2007, following the recrafting of the Lisbon competitiveness agenda to focus more narrowly on growth and job creation, the Competitiveness Council and the Commission launched an Action Program to reduce the administrative burden of EU regulation. The program contains numerical targets for administrative cost reductions and led to creation of an Impact Assessment Board within the Commission that is independent of policy-making departments, as well as an advisory "High Level Group of Independent Stakeholders on Administrative Burden" (known as the "Stoiber Group" after the group's appointed chair, Edmund Stoiber, former minister-president of the German state of Bavaria).

The Stoiber Group was given an initial three-year mandate and assigned to provide the Commission with advice on measures to reduce administrative burdens following consultation with economic actors; in August 2010, the Commission extended the group's mandate through 2012.[21] The group serves as an aggregation point for suggestions for reducing administrative cost of regulations from firms, national and European industry associations, and public authorities. As part of the European Commission's "Action Plan for Reducing Administrative Burdens,"[22] the Commission is expected to present to the Competitiveness Council its responses to these suggestions aggregated by the Stoiber Group.

The EU's "better regulation" agenda has progressively come to focus on minimizing the regulatory cost burden on industry, motivated by the logic that only a more efficient regulatory environment would make it possible for the EU to attain social and environmental objectives "without disproportionate administrative costs" (Competitiveness Council March 7, 2005). The result has been a steady extension of regulatory impact assessment not only to pending legislation, but also retroactively to legislation already adopted. The College of European Commissioners that came to office in 2004 inaugurated a process of reviewing all legislation adopted under its predecessor Commission, using as the standard for assessment the compatibility of regulations with the Lisbon agenda's attention to growth and jobs (Commission, 2006: 7). In response to the assessment, the Commission withdrew a substantial number of proposals for new regulation (Commission, 2006; 2008).

In addition to more extensive reliance on regulatory impact assessment, the Competitiveness Council and the Stoiber Group have pressed for substantive shifts in the impact assessment process involving earlier and more frequent consultation of industry and greater reliance on quantification of costs. The Competitiveness Council has asked for industry consultation and impact assessment to occur earlier in the policy-development process, in order to allow the Commission to consider alternatives to regulation before proposals are tabled (Commission, 2008: 4).[23]

Following its creation in 2002, the Competitiveness Council took on the comprehensive chemicals sector regulation (REACH) proposed by the Commission as its first substantive issue. The Council's objective was to ensure that efforts to protect public health and the environment did not impose excessive costs on chemicals producers.[24] As detailed in chapter 2, the Council, along with the European Parliament's IMCO, was instrumental in transforming a proposal that initially bore a strong imprimatur of environmental and health interests by significantly reducing the regulatory cost burden on industry (Smith, 2010).

In the case of REACH, the Parliament's IMCO served as a crucial point of legislative access for chemical-industry advocacy. Industry interests succeeded in neutralizing a strong environment and health lobby within the EP's Environment Committee. As the committee chiefly responsible for coordinating the position of the EP, the Environment Committee had to incorporate critical elements of the IMCO's position (having in particular to do with requirements for registration of dangerous chemicals) into its report on the proposed regulation in order to garner majority support from the full EP. As demonstrated in chapter 2, the European-level chemicals-industry federation was able to modify the EP's position by working through its relationships with members of the European Parliament (MEPs) in the Internal Market Committee rather than having to compete for influence with environmentalists operating within the Environment Committee's network. Industry's impact was evident in the outcome of the Parliament's first-reading position, which departed sharply from the heavily pro-environmentalist tenor of the European Commission's original proposal.

In the case of health claims on food labels, food producers similarly utilized the Parliament's IMCO as a source of leverage. Industry advocates again moved the EP toward a sympathetic first-reading position; in this instance, the Parliament removed the provision for nutrition profiles—barring health claims in instances in which unhealthful components were present in foods in substantial amounts—that was a centerpiece of the originally proposed legislation. As analyzed in chapter 3, the critical difference between chemicals and health claims on foods was that in the latter case, national health ministers were determined to reclaim regulatory control. When the Council restored the nutrition profiles provision in its common position on the regulatory proposal, this altered the balance of forces in the EP in its second reading, and the Council's position ultimately prevailed.

In sum, producer interests have adapted effectively both to the expanded regulatory remit of the community and the movement toward weighted majority decision making in the Council of Ministers that has reduced the ability of individual governments to control policy outcomes. Nonetheless, regulatory rigor has intensified in the EU context in the cases of chemicals, health claims on food labels, and recycling requirements for end-of-life vehicles and other e-waste—in sharp contrast with outcomes in the United States.

## ORGANIZATION AND CHAPTERS

The cases of chemicals, health claims on food labels, and ELV recycling demonstrate that environmental and health regulation in the EU is distinguished from that in the United States not simply by degree of rigor, but also by a trend toward compulsory regulation. While there is a good deal of overlap in the extent to which cost-benefit analysis and implications for economic competitiveness shape the regulatory environment in the EU and the United States, the tendency toward compulsory regulation reflects the integration objective embedded in the EU's regulatory trilemma.

Chapter 2 illustrates the argument with respect to chemicals regulation. In this sector, the EU recently has introduced a compulsory regulatory regime, while a patchwork of voluntary industry initiatives and industry-government partnerships characterize the regulatory landscape in the United States. At the same time, the EU's regulatory regime, which bore the strong imprimatur of European environmental NGOs in the legislative drafting stage, was revised in its final version in accordance with many of the objections of industry interests—resulting in less restrictive provisions for authorization and registration of potentially hazardous chemicals.

Chapter 3 turns to health claims on food labels. This chapter again illustrates the regulatory differences that emerge from the effort to reconcile competitiveness and health objectives in the United States versus the struggle to balance competitiveness, public health, *and* advancement of European integration in the EU. While court rulings based on the principles of proportionality of regulatory measures and free commercial speech have unleashed a health claims free-for-all in the United States, EU institutions have responded to a welter of health claims and the diversity of national labeling regulations by introducing a more restrictive EU-wide regulatory regime. Regulatory rigor is not a product of the absence of industry mobilization and lobbying weight—industry associations both sought European-level regulation to establish legal certainty and exercised considerable leverage over the position of the EP, especially in its first reading of the regulation initially proposed by the European Commission—but instead reflects the determination of national health ministers to recapture at European level some of the regulatory control lost through the impact of ECJ decisions on national regulation.

In chapter 4, I examine regulation of recycling of vehicles reaching the end of their useful lives. ELV recycling reproduces regulatory patterns evident in recycling of other end-of-life products, including electronic waste (computers) and appliances. While the US regulatory regime emphasizes "good product stewardship," essentially a call for good public citizenship by producers and shared responsibility for waste reduction and recycling across those involved in the product life cycle, the EU has instituted a mandatory regulatory regime based on the "producer pays" principle. While compliance with the recent EU regulation remains poor, the European Commission has tolerated a relatively high compliance deficit because it was able to capture through European-level regulation substantial benefits in the form of European-level interest aggregation and articulation, including stronger coordination by producer and environmental lobbies.

The concluding analysis in chapter 5 reinforces the similarity of fundamental elements of the EU and the US regulatory contexts. Industrial competitiveness is a central concern in both cases, and policy making draws heavily on regulatory impact assessment. Courts uphold strict proportionality in the application of regulatory remedies and give extensive protection to free commercial speech in both venues. Ultimately, the third leg of the EU's regulatory trade-off—the objective and process of European integration, including the mobilization of national government ministers at EU level to achieve regulatory objectives—is critical to explaining the divergence of EU and US regulation to protect public and planet.

CHAPTER 2

# Chemicals Regulation: Courts Rule in the United States; Industry Ascendant in the EU

THIS CHAPTER DEMONSTRATES THREE ELEMENTS OF THE BOOK'S core argument. First, despite the stark current differences in regulatory regimes, the principle of precaution was central to the origins of contemporary chemicals regulation not only in the European Union (EU), but in the United States as well. Second, court decisions imposing extraordinary demands on regulatory authority have been a pivotal element of the erosion of regulatory rigor in the United States. Third, while EU institutional structures provide inviting points of access for organized environmental interests, organized industry interests have countered this access; as a result, a rising focus on the burden of regulatory costs on industrial competitiveness has tempered the intensified rigor of EU chemicals regulation. In short, while outcomes diverge, the broader ideational and political landscape undergirding regulatory policy differs much more subtly between the United States and the EU than most observers recognize.

The chapter traces the regulatory regime emerging from the 1976 Toxic Substances Control Act (TSCA) in the United States and the trajectory leading to the EU's 2006 REACH (Registration, Evaluation and Authorization of Chemicals) regulation. As the chapter demonstrates, chemicals regulation in the United States follows an arc from a regulatory regime based on the precautionary principle through a series of court cases establishing evidentiary standards and cost-benefit analyses of regulatory remedies that eviscerate the authority of the Environmental Protection Agency (EPA) and undermine regulatory rigor. Along the way, chemicals manufacturers have preempted revisions to TSCA by adopting voluntary regulations focused largely on improving public perceptions of the behavior of chemicals producers. In contrast, the EU has moved from a patchwork of national regulatory systems, including voluntary schemes, through a regulatory process initially shaped by trans-European environmental interests but ultimately moderated by the force of organized industry interests and pervasive concerns with industrial competitiveness—though still substantially exceeding US regulatory standards.

The chapter first discusses the regulatory regime for chemicals in the United States. The discussion explains why the 1976 TSCA has strayed so far from its precautionary roots, with a particular focus on the role of courts. The chapter then examines the EU's REACH legislation, beginning with its origin in the environmental movement and the support for environmental interests within EU institutions. The discussion next addresses the surging influence of organized industry interests in the EU's regulatory policy-making process. The final section highlights continuing elements of support for a precautionary approach to chemical regulation in the United States, evident in ongoing debates about reforming TSCA.

## Precaution and the Toxic Substances Control Act

The TSCA was part of the stream of environmental legislation produced by the US Congress in the 1970s and the 1980s.

In contrast with environmental laws such as the Clean Air and Clean Water acts, TSCA is designed to regulate hazardous chemicals at the production and distribution stages rather than to address releases of chemicals into the environment. TSCA itself grew out of a 1971 report by the Council on Environmental Quality (CEQ), an executive agency created by the 1969 National Environmental Policy Act. In a climate of heightened public attention to the dangers of chemical exposure associated with media accounts of health threats arising from exposure to mercury, vinyl chloride, arsenic, and asbestos (Reynolds, 1977: 54; Haemer, 1999: 108–109), the CEQ articulated a precautionary approach to chemicals control, urging that "[w]e need no longer be limited to repairing the damage after it has been done" (Reynolds, 1977: 39, note 12; Ruggerio, 1989: 81). As a researcher from the Natural Resources Defense Council wrote optimistically in the immediate aftermath of TSCA, "By granting EPA the authority to act on chemical categories, Congress provided the Agency with an unprecedented opportunity to break away from past methods of controlling chemicals" (Slesin, 1978).

Indeed, TSCA requires that companies notify the EPA prior to production of new chemicals. This premanufacture notice (PMN) provision of TSCA calls for producers to furnish the EPA with basic data on each chemical—including production process and volume, intended uses, exposure and release levels, and information about disposal and by-products—at least 90 days prior to production or import. Section 5 of TSCA establishes conditions under which the EPA has authority to penalize firms for nonnotification or to regulate a substance by limiting or prohibiting production or distribution (Ruggerio, 1989: 86; Hanan, 1992: 403–404). Section 4 mandates that the EPA Administrator demand additional testing of chemicals that "may present an unreasonable risk of injury to health or the environment," while Section 6 authorizes the EPA to regulate chemicals already on the market if there is a "reasonable basis" to conclude that they pose such an "unreasonable risk" (Reynolds, 1977: 75). The EPA may regulate the manufacture and distribution of a chemical through means ranging from mandatory warning labels to precautions that must be taken in manufacture, use, or disposal of the chemical, to outright prohibition (Gaynor, 1977: 1151; Ruggerio, 1989: 87; Hanan, 1992: 404). In formal terms then, TSCA extends substantial regulatory authority to the EPA to demand information, require testing, and act against suspected risks.

In his legal analysis in the immediate aftermath of the passage of toxic substances control, Kevin Gaynor argues that in legislating TSCA, Congress arrived at the "reasonable basis" criterion because it was aware that scientific certainty may be elusive in the assessment of the toxological effects of a chemical substance (Gaynor, 1977: 1154); the result is a permissive environment for regulatory action by the EPA—due to the likelihood of inadequate data, "Congress went to great lengths to explain that under TSCA the Administrator could overcome the problem through reasoned speculation and extrapolation from existing data" (Gaynor, 1977: 1161). Furthermore, in reviewing EPA actions, courts would acknowledge the limits of available evidence and make similar allowance for uncertainty by deferring to the judgment of the EPA Administrator (Gaynor, 1977: 1161).

Despite the precautionary language of TSCA and the expectation of broad regulatory leverage,[1] it is almost universally judged ineffective by observers (Ruggerio, 1989; Hanan, 1992; GAO, 1994; Percival, 1998; Haemer, 1999).[2] Few chemicals actually have been regulated by the EPA under the provisions of TSCA (Hanan, 1992: 396). Of 24,000 premanufacture notices submitted to the EPA between 1976 and 1994, approximately 90 percent were approved without restriction or any requests for additional test data.[3] Furthermore, the EPA has even thinner information and exercises less regulatory control over the chemicals that were already in commerce when TSCA became law; the 62,000 chemicals in commerce as of 1979 are exempt from TSCA's premanufacture notice provision.[4]

The EPA has attempted to use Section 6 to impose controls on only five new chemicals or groups of chemicals since 1979[5]; the majority of these efforts have been unsuccessful (Guth, Denison, and Sass, 2005: 7; Lowell Center, 2003: 3; GAO, 2005). The EPA's effort to regulate asbestos during the 1980s indicates that even where there is substantial accumulated evidence of harmful impact on health, widespread use, and characteristics of the substance increasing the potential for high levels of exposure, TSCA gives the EPA little leverage to act. Following a decade of careful consideration by the EPA, a 1990 decision by the US Court of Appeals for the Fifth Circuit rejected the EPA's ban on asbestos. In general, EPA regulation under Section 6 rulemaking has been so rare that Congress has had to intervene several times to regulate specific hazards—including asbestos in schools in 1986, indoor radon gas in 1988, and lead paint in 1992 (Haemer, 1999: 119). The EPA has not invoked its toughest regulatory sanctions since it banned dioxin in 1979.

Why has TSCA been so ineffective? First, because TSCA fundamentally departs from the other major pieces of environmental protection legislation of the 1970s, such as the Clean Air Act and Clean Water Act, in that it seeks to regulate inputs into the production process rather than outputs (Hanan, 1992; Guth, Denison, and Sass, 2005), implementation is especially difficult. Additionally, the EPA is severely handicapped in its efforts to garner sufficient information about chemicals to invoke TSCA's regulatory provisions. Finally, courts have established a prohibitively high judicial standard for regulatory action.

The EPA is swamped with premanufacture notifications (PMNs); chemical producers are not required to submit toxicity data, and half of all PMNs provide none. The EPA has a mere 90 days to screen chemicals for risk and faces a high hurdle for requiring testing. A 1998 EPA study revealed that the basic "Screening Information Data Set" established by the OECD—a set of preliminary screening data—was publicly available for just 7 percent of the chemicals produced in volumes exceeding 1 million pounds per year in the United States (Guth, Denison, and Sass, 2005: 3).

As noted, the EPA may regulate substances when there is a "reasonable basis" of an "unreasonable risk," language that lays a foundation for the exercise of precaution in implementing TSCA. However, in their interpretation of this language, courts typically have required EPA to submit toxicity data in order to meet the evidentiary standard for the demand of further testing. Since such data typically is not provided by manufacturers, the ability of the EPA to invoke its authority to require additional testing is severely constrained (Guth, Denison, and Sass, 2005: 4; 6). Where the EPA has sought to require testing, its efforts typically have been challenged by chemical manufacturers. Courts have established a high threshold for EPA test rules, which withstand judicial scrutiny only when the EPA has a "substantial" basis for suspecting an unreasonable risk—meaning that the risk must be more than theoretical and the level of toxicity of the substance high (Bergeson, Campbell, and Rothenberg, 2000: 10–11).[6] As Bergeson et al. indicate, "Issuing Section 4 test rules has proven to be exceedingly time consuming, resource intensive, and thus costly" (Bergeson, Campbell, and Rothenberg, 2000: 11). Over a quarter century of regulation, the EPA has been able to induce further testing for only about 200 chemicals (Guth, Denison, and Sass, 2005: 4; GAO, 2005).

Even where the EPA cannot compel testing, producers are required to submit existing toxological studies. However, compliance levels are relatively low. A onetime compliance audit conducted by the EPA beginning in 1991 revealed 11,000 unreported studies on potentially hazardous chemicals. Out of 123 companies audited, 89 had information that should have been forwarded to the EPA but was not.[7] Indeed, most of the EPA's regulatory enforcement capacity is absorbed by action against companies that have failed to file reports (Haemer, 1999: 114).

Finally, when the EPA does exercise its authority to regulate a chemical, it must impose the least burdensome restriction. Courts have called for cost-benefit justifications weighing all possible levels of restriction of a particular chemical as well as the costs of alternatives. Ultimately, the legal structure of TSCA "presumes that manufacturers have the right to market chemicals and places a heavy burden on government to prove the need for regulation before it can interfere with that right" (Guth, Denison, and Sass, 2005: 3). The nearly insurmountable standard for judicial review reflects the ambiguity of priorities under TSCA—on the one hand, the regulation refers to the necessity of testing and places the burden on those who manufacture and process chemical substances; on the other hand, the law stipulates that regulation should not create "unnecessary economic barriers" (Hanan, 1992: 410). As Government Accountability Office (GAO) studies of TSCA implementation indicate, EPA officials believe they are hampered in their efforts to regulate potentially toxic chemicals by difficulties in documenting that EPA restrictions on a chemical represent the least burdensome regulatory option (GAO, 2009a: 10).

In the 1990 asbestos case, known as "Corrosion Proof Fittings," the court ruled that the hurdle for the EPA to establish "substantial evidence" for its claims regarding the dangers of asbestos was especially high because it was seeking to ban the substance—the harshest remedy available under TSCA (Hanan, 1992: 413; Davies and Mazurek, 1998: 24). Lettie Wenner classifies the Corrosion Proof Fittings decision as "[a]n extreme example of judicial intervention into administrative decisionmaking."[8] As Andrew Hanan asserts, "[T]o require specific cost-benefit calculations with respect to all potential risks, all possible regulatory alternatives, and the myriad of potential product substitutes virtually negates meaningful chemical regulation" (Hanan, 1992: 416).[9]

Robert Haemer, noting that the procedural requirements incorporated into TSCA were designed to balance the EPA's broad powers to identify and regulate risks with an economic calculus, also argues that judicial interpretation has distorted the regulatory process. "Rather than establish balance," he argues, "these procedural mechanisms have allowed industry to strike down or delay many of the rules promulgated by EPA" (Haemer, 1999: 106). The expressed views of EPA officials confirm this perspective. In testimony before the US Senate's Subcommittee on Toxic Substances, Research and Development in 1994, the EPA's Director for Environmental Protection Issues, Peter Guerrero, attributed TSCA's ineffectiveness to legal standards, noting that requirements "for taking regulatory action are so high that EPA has been discouraged from attempting to regulate chemicals and has given implementation of the act low priority" (GAO, 1994: 2). Furthermore, in his testimony, Guerrero contrasted TSCA with Canada's Environmental Protection Act of 1988, pointing out that the Canadian law provides for control of chemicals harmful to the environment or human health without regard to the cost-benefit calculus of control; the cost-benefit calculus only enters into the decision concerning the type of regulatory control imposed (GAO, 1994: 7).

In his review of outcomes of appeals of EPA environmental rulings filed by business interests in federal courts, Sheldon Kamieniecki finds that business and the EPA each succeed with about equal frequency (Kamieniecki, 2006). However, the EPA has a much higher success rate sustaining rulings involving the Clean Air Act than it does for TSCA.[10] Overall, Kamieniecki finds that business has a much greater impact over issue framing and court interpretations of the application of regulations than over agency rulemaking itself (Kamieniecki, 2006: 140; 254).

Finally, in their examination of the development of federal judicial review of EPA actions, Glicksman and Schroeder argue that decisions in the first decade or so after the creation of the EPA bore fundamental features favorable to environmental interests. Based on the view that legislation emerged from careful efforts by Congress to balance multiple interests with clear public policy objectives, courts sought to rule on contested administrative decisions by discerning legislative intent; they also developed by the 1970s a philosophy grounded in a concern with the dangers of regulatory capture by industry and committed to the notion that environmental quality is a privileged and consensual public interest (Glicksman and Schroeder, 1991: 271). As a consequence, "[A]gency decisions disrespectful of environmental interests were more likely to be subjected to intense judicial review and decisions in which the agency had acted to protect the environment or public health under conditions of uncertainty were more likely to receive more deferential treatment" (Glicksman and Schroeder, 1991: 273).

However, this favorable climate for environmental concerns did not endure. By the end of the 1970s, courts, informed by the evolution of public choice theory, began to act instead on the perception that individual legislators may be motivated by self-interest rather than the public interest. In this regard, legislators might well act to simultaneously satisfy conflicting constituencies by adopting legislation to address the concerns of environmental interests, while establishing conditions for lax implementation to satisfy opponents of regulation. In this context, legislation becomes substantially symbolic rather than demarcating a firm public policy commitment (Glicksman and Schroeder, 1991: 294). In addition, courts no longer accorded environmental interests an exalted status, viewing environmental concerns instead as one among many competing values. Thus, "In the 1970s, and more so in the 1980s, more and more arguments over environmental policy advanced an economic understanding of environmental values" (Glicksman and Schroeder, 1991: 282).

This movement toward the treatment of environmental issues as "ordinary" invited the application of cost-benefit analysis to decisions on environmental regulation (Glicksman and Schroeder, 1991: 284). Judges, in other words, came to view environmental rules as the product of political bargains that should not be disturbed by the judicial branch. Since regulations represent resolutions of disputes between competing values rather than clear statements of policy, courts should defer to the technical judgments of regulatory agencies like the EPA. Nonetheless, consistent with the application of cost-benefit criteria, courts should intervene to ensure "reasoned" application of such judgments. As Glicksman and Schroeder point out, courts have not hesitated to judge EPA actions arbitrary (Glicksman and Schroeder, 1991: 296).

Two critical questions emerge from this perspective on the Toxic Substances Control Act and its development from precautionary legislation to ineffectual regulatory regime. First, how did industry respond to original efforts to regulate chemicals, and what role did they play in determining the contours of the TSCA legislation passed in 1976? And second, why, given the massive gaps in regulatory effectiveness, has there been no updating of TSCA to fill in these shortcomings? The question is particularly acute given that TSCA "is the only U.S. pollution law that has never been modernized or strengthened since original passage" (Environmental Working Group, 2009b).

## TSCA and Chemicals Industry Lobbying

Chemicals manufacturers lobbied heavily in an effort to mold TSCA to their preferences if they could not defeat regulation of chemicals production outright.[11] In particular, organized industry interests secured effective legislative access through the House of Representatives, where members introduced industry-friendly bills or advanced amendments to reduce the scope of toxic substances control and the reach of EPA administrative authority. At the same time, industry did not succeed in its effort to forge a producers coalition with organized labor, since the unions tended to seek to address concerns of workers about protection from chemical exposure (Reynolds, 1977: 56). In addition, the high salience of environmental and health threats and public perceptions about rising dangers from toxic chemicals inclined many legislators toward adopting legislation to control chemical hazards. Ultimately, industry influence meant that critical aspects of the significance of the legislation would be determined by court proceedings, but not necessarily that regulatory effectiveness was undermined from the outset. This explains why, as discussed above, numerous early assessments by scholars and practitioners depicted TSCA as a breakthrough in regulatory rigor.

In their comparison of chemicals regulation in Britain, France, Germany, and the United States, Brickman et al. emphasize the role of autonomy of the legislature from the executive (nominally higher in France than in Britain or Germany, but clearly highest in the United States), and, especially, the significance of contention, competition, and fragmentation *within* the legislature as institutional features determining the nature of regulation. In parliamentary systems, drafting of legislation typically takes place under executive authority, with involvement of the legislature coming much later in the legislative process (Brickman, Jasanoff, and Ilgen, 1985: 63). As Brickman et al. suggest, the consequence of early legislative involvement is a protracted struggle to meet the demands of competing interests, and the result is "more an elaborate system of procedural checks on administrative discretion than a genuine bipartisan agreement on a coherent set of policy prescriptions" (Brickman, Jasanoff, and Ilgen, 1985: 63).

Although chemicals regulation rose to the top of the policy agenda with the report of the Council on Environmental Quality in 1971, legislation was delayed through two Congresses as environmental and industry interests penetrated the Senate and the House, and contending versions of a new regulation produced deadlock. While Senate and House versions of the proposed legislation differed from the beginning of the debate, ultimately the legislation became bogged down within the House. Inability to agree on a version of the bill in the House precluded convening of a House-Senate conference committee prior to legislative adjournment of the 92nd Congress in January 1973 (Reynolds, 1977: 41). Even after inaction through the 92nd and 93rd Congresses, multiple versions of the bill competed in the House during the 94th Congress. Issues of contention included the extent of required premarket screening of chemicals and the scope of authority granted the EPA Administrator; the irreconcilability of Senate and House positions on these issues substantially reflected the influence of chemicals industry representatives in the House (Reynolds, 1977: 40–42).

Penetration of chemicals sector interests in the House also was reflected in the wider diversity of bills introduced there, including at least one bill proposing an extremely permissive regulatory environment (Reynolds, 1977: 43). Amendments proposed on the House floor during the 92nd Congress testified to the access of industry interests; these included a broadening of exemptions from premarket screening of substances and a provision requiring the EPA Administrator to submit a detailed economic impact statement when promulgating a rule under the regulation (Reynolds, 1977: 41, fn 21). Later, in the 94th Congress that ultimately produced TSCA, the House favored substantial restraints on the power of the EPA Administrator to keep new chemicals off the market following notification; while the Senate version would have permitted the EPA to promulgate a new rule following administrative hearings and a 90-day notice period, the much more administratively restrictive House version required the EPA Administrator to seek a court injunction (Reynolds, 1977: 49).

The bill that ultimately emerged from a House-Senate conference committee bore a heavy symbolic dimension, reflecting a genuine desire shared by many legislators to act to protect human health from the potential dangers of toxic substances, tempered by hesitation to impose a costly regulatory burden on industry (Rogers, 1988; Glicksman and Schroeder, 1991; Percival, 1998: 14). Indeed, as Reynolds reports, when Representative

Robert Eckhardt introduced the House bill that was to be the basis of the House compromise with the Senate, he justified the bill as meeting the need "for an effective means of controlling toxic chemicals . . . (without) an undue regulatory burden placed on the chemical industry" (Reynolds, 1977: 46).

Industry lobbying was intense, particularly since there were advocates in Congress who favored a precautionary approach (Reynolds, 1977: 50, esp. note 51). In the original debate over TSCA, lobbying from the chemicals industry focused on a core objective of preventing any mandatory testing of chemicals already in use (Environmental Working Group, 2009b). But industry positions varied, from outright opposition to regulation by some large as well as small producers, to the alternative strategy of the Manufacturing Chemists Association (from 1978, the Chemical Manufacturers' Association [CMA], then, beginning in 2000, the American Chemistry Council), which took the approach of supporting a limited bill predicated on careful cost-benefit analysis (Reynolds, 1977: 50; 53). Dow Chemical, repudiating the need for regulation, engaged in the tactic that European chemicals producers later invoked in the debate over REACH, projecting and publicizing exorbitant cost estimates and warning of dire consequences for employment and innovation in the sector (Reynolds, 1977: 52–53). Employing another tactic later echoed in the EU's REACH debate, chemicals producers warned of how the costs of regulation would impede innovation.[12]

## TSCA and Regulatory Preemption

What has prevented a tightening of the regulatory regime for chemicals in response to growing recognition of the ineffectiveness of TSCA in protecting safety and health? The absence of additional regulatory action to amend TSCA is largely a story of industry preemption of stricter regulation through voluntary programs. Thomas Lyon and John Maxwell divide voluntary approaches to environmental regulation into three subtypes: unilateral commitments, consisting of business-initiated programs; public voluntary schemes, which involve agreement of participating firms to implement standards developed by public bodies; and negotiated agreements, which emerge from government-industry dialogue (Lyon and Maxwell, 1999). These authors find that unilateral commitments are more common in the United States, while the latter two forms predominate in Europe. We may classify unilateral commitments as a form of regulatory preemption, the primary motive of which is to allow industry to control the regulatory environment and forestall the imposition of more extensive and rigorous regulation.

On the one hand, important strands of business management theory tend to contradict the notion of preemption; the "Porter hypothesis," for example, indicates that businesses may operate for prolonged periods under routinized procedures that leave unmined possibilities for efficiency gains through innovation that improves environmental outcomes (Porter, 1995). Regulation is therefore a means to shake firms out of routine and induce them to capture available gains. However, the prevailing perspective of regulatory economics is that such untapped benefits are modest; inefficiencies exist but are hardly pervasive, suggesting that regulation is in fact costly to firms,[13] and voluntary measures likely represent "a rational readjustment response to new regulatory costs" (Andrews, 1998: 180; see also Palmer, Oates, and Portney, 1995).

Indeed, John Maxwell, Thomas Lyon, and Steven Hackett (2000) find that self-regulation may be preemptive: firms engage in more self-regulation as they perceive a rising threat of compulsory government regulation. These authors construct a model in which the likelihood of preemptive self-regulation depends upon the costs of coordinated action for industry and the scope of collective action problems faced by citizens potentially willing to engage in environmental activism. Where collective action problems are significant and firms face low coordination costs (typically the case where the number of firms is relatively small), the modest environmental benefits of self-regulation will suffice to induce citizens to refrain from action to influence the policy-making process. For producers, the impetus to preemptive self-regulation is reinforced where environmental damage is high, rendering the net costs of sufficient abatement to undermine collective action by citizens relatively low (Maxwell, Lyon, and Hackett, 2000).

The Chemical Manufacturers Association's "Responsible Care" program, ostensibly designed to reduce chemical hazards, is perhaps the most significant unilateral commitment and instance of preemption in the chemicals sector. The CMA introduced Responsible Care in 1988; most sources trace the genesis of the program to the 1984 Union Carbide disaster in Bhopal, India; the adoption in Canada of a program to publicize industry efforts to improve their environmental performance (Lyon and Maxwell, 1999: 3–4)[14]; survey evidence of waning public confidence in chemicals safety in the United States (King and Lenox, 2000: 699); and the creation by the US Congress in 1986 of the Toxic Release Inventory (TRI), requiring US manufacturers to publicly report toxic emissions (Sissell, 2008). Rees suggests the rise of environmentalism, attendant growing public awareness of toxic waste discharges, and the Bhopal event produced a legitimation crisis for the US chemicals industry that demanded a strategic response (Rees, 1997: 484–486). The CMA—later the American Chemistry Council (ACC)—represents approximately 180 firms that account for the bulk of the chemicals volume produced in the United States; the CMA describes its principal function as protecting industry from external intervention (King and Lenox, 2000: 699; Rees, 1997).

With such a large membership, the CMA would appear to face substantial coordination costs, a potential obstacle to unilateral self-regulation. However, the CMA resolved coordination problems by requiring all member firms to adopt Responsible Care. The leadership of the industry's large chemicals producers—and their leverage over the many smaller producers they supply—facilitated this approach (Andrews, 1998: 181). The emphasis on information provision—the first of ten "Guiding Principles" of Responsible Care is to "[r]ecognize and respond to community concerns about chemicals and our operations"—and safety effort (as opposed to outcomes) attests to the extent to which the objective of Responsible Care was as much to affect public perception as to improve environmental performance of the chemicals sector. Indeed, community advisory panels (CAPs) designed to foster industry dialogue with publics neighboring chemical plants are a central part of the Responsible Care program.[15]

As Michael Givel writes, the pilot program established at the inception of Responsible Care involved extensive advertising, polling, and use of focus groups designed to improve the industry's communications campaign.[16] And an internal CMA document identifies forestalling public calls for additional regulation as a primary motive for the program (Givel, 2007: 89). The editor-in-chief of *Chemical Week*, the industry magazine for the chemicals sector, delineated the motive for Responsible Care in 1999, when he wrote: "One of the reasons why Responsible Care was adopted in the U.S. 11 years ago was to help industry get ahead of command and control regulations. The driver was the belief that taking the initiative on improved environmental performance was the best way to preempt the industry being stifled by ever more onerous regulations piled upon it."[17] Based on evidence they gathered from industry observers, Peter Simmons and Brian Wynne echo this finding that a central objective of Responsible Care was "limiting state intervention to a level that is acceptable to the industry" (Simmons and Wynne, 1993: 205).

The process for implementing Responsible Care is self-assessment at the firm level; there is no third-party verification or enforcement of standards (King and Lenox, 2000: 700).[18] Self-assessment can in theory be productive where the spread of norms and values transmit best practices across firms. However, Andrew A. King and Michael J. Lenox find evidence of substantial opportunism—many firms in the Responsible Care program are "poor performers," and toxic emissions in fact decline more slowly among program participants than among nonmembers. Furthermore, firms with high levels of production and above-average levels of pollution were more likely to be members of CMA and to participate in Responsible Care.[19] King and Lenox conclude that the program has been geared not toward protecting the environment, but toward protecting the "reputational commons" of the chemicals industry and CMA membership (King and Lenox, 2000: 713). And Joseph Rees argues that the communitarian nature of the Responsible Care program reflects the role adopted by the CMA as a trade association in the wake of the TSCA, which constitutes a regulatory community without any real authority, substantively subordinate to member firms that implement agreed industry policies as they see fit (Rees, 1997: 507).[20]

Preemptive self-regulation has been effective largely because the EPA's limited resources, and the structure of TSCA generate EPA dependency on industry for data. The EPA cannot compel industry testing of a chemical based on either suspicion of potential toxicity or large production volume alone. As US GAO studies of TSCA implementation show, the process by which the EPA must develop a basis for requiring additional testing of a substance, consider public comment of its proposed rule, and promulgate a rule requiring testing is costly and takes from two to ten years (GAO, 2009a: 5). The EPA simply cannot undertake this process for all substances for which there is suspicion of potential harm to human health and the environment.

Closely related to the problem of limited EPA resources, political tensions inherent in the regulatory dilemma involving trade-offs between environmental and health objectives, on the one hand, and regulatory costs, on the other, also favor acceptance of self-regulation. As Rena I. Steinzor asserts, "Demands that EPA replace traditional command and control regulation with market-based alternatives are uniformly embraced by key members of Congress and regulated industries, who argue that the same level of protection could be achieved at much lower costs" (Steinzor, 1998: 200). This logic explains why EPA has extensively sponsored industry self-regulation projects, many with dubious environmental benefits. Faced with a regulatory dilemma and a contentious political environment along with the inherent uncertainties of scientific data, the cost to the EPA of outcomes that tilt suboptimally (from an environmental and health perspective) in favor of industry is much lower than for European Union institutions that also seek to use the regulatory process to advance the cause of European integration.

## Environmentalists and REACH in the EU

In the case of the EU, international environmental initiatives during the 1990s contributed to the impetus for development of a new regulatory regime for chemicals at the European level. Throughout the 1990s, international environmental activism drew mounting attention to dangers to human health resulting from bioaccumulative substances. Among other evidence, these included a Swedish study documenting rising levels of toxic fire retardants in breast milk.[21] By spring 1998, these health concerns reached the agenda of environment ministers from Austria, Denmark, Finland, the Netherlands, and Sweden, who ultimately induced the EU's Council of Environment Ministers to ask the European Commission to review existing chemicals sector legislation (Shörling, 2004: 55–56).

The review of chemicals regulation conducted by the Commission identified grave shortcomings in the existing regulatory environment. The central problem was the inadequacy of requirements for the gathering of safety data and risk assessment, including the complete exclusion from any such assessment for substances marketed prior to September 1981. In response to these findings, the Council asked the Commission to develop a new regulatory framework for the sector based on the precautionary principle. This entailed: (1) requiring that firms provide safety data before gaining authorization to bring chemicals to market in cases where there is evidence or suspicion of a danger to health and environment; (2) a shift in the burden of proof for demonstrating safety from public authorities to industry; (3) a single system for the review of all chemical substances, including those on the market prior to 1981; and (4) incentives for technical innovation (RAPID IP/03/1477, October 29, 2003; Shörling, 2004: 59 ; 62).

In sum, the legislative process began with regulatory objectives defined by environmental and health concerns. Environmental NGOs called for a regime that would fully identify risks, provide the public with information about the hazards of chemicals in the environment and in the products they consume, and establish a process for completely phasing out the production of harmful chemicals by a fixed date. In an atmosphere of heightened awareness of the dangers posed by chemicals production, organized environmental interests found support in the European Commission's Environment Directorate General and in the European Parliament's (EP) Environment Committee (ENVI), which was responsible for writing the EP's position. Indeed, a broad, diverse coalition developed in support of a new regulatory framework focused on a precautionary approach to the approval of chemical products. The coalition included a range of environmental NGOs, numerous health advocacy organizations, and the European consumers' organization, Bureau Européen des Unions de Consommateurs (BEUC).[22]

In April 2003, the Swedish Society for Nature Conservation, WWF Sweden, Fältbiologerna, and Friends of the Earth Sweden founded the International Chemical Secretariat (ChemSec) to coordinate information and lobbying activities regarding REACH, the European Commission's draft legislation regarding registration, evaluation, and analysis of chemicals. A group of large retailers sharing an interest in bolstering consumer trust, preventing further "toxic scares" such as those posed by Mad Cow disease and dioxins in food, and obtaining clear information about chemical content from suppliers—including the clothier H&M and Britain's Marks & Spencer—organized through ChemSec in support of a rigorous regulatory approach.[23] This environmental coalition gained access to the EU policy-making process via the Council of Environment Ministers, the Commission's DG Environment, and the Environment Committee of the European Parliament. The ENVI, led by a Green Party rapporteur, shaped a response by the Parliament to the 2001 European Commission white paper on chemicals regulation that called for even stronger health and environmental safeguards than those suggested by the European Commission.

The European Commission presented its formal regulatory proposal in October 2003. With the European Parliament's ENVI designated to serve as the responsible committee for drafting the position of the EP, the environmentally friendly character of REACH seemed assured. It was likely from the outset that concessions to industry concerns about the costs of implementing the regulation and its impact on the international competitiveness of Europe's chemicals sector would prove necessary. However, it was at this point in the policy-making process—as the EP undertook its first reading of the formal legislative proposal and the Council of Ministers representing member state governments began negotiating toward a common position—that industry advocacy and a discourse focused on the regulatory costs accruing to industry had a dramatic effect within both the Council of Ministers and the Parliament. An intensification of industry organization and lobbying was accompanied by institutional developments that accommodated industry interests. These included the formation of a new configuration of the Council of Ministers focused on industrial competitiveness and, within the EP, the emergence of the

Internal Market Committee (IMCO) as an advocate for industry concerns, a development that counterbalanced the access of environmental interests to the Parliament's Environment Committee.

## IMPACT ASSESSMENT

The course of REACH represented a break through in codifying the precautionary principle in chemicals regulation in the EU. There is little indication that the impact of REACH will replicate the impotence of the TSCA in the United States. Nonetheless, the final version of the regulatory framework involved significant reductions in information provision and requirements for substitution of dangerous chemicals, leaving environmentalists disappointed.[24] Balancing the points on the triangle representing the EU's regulatory trilemma, EU institutions advanced the objective of European integration by developing a comprehensive regulatory framework at the European level whose rigor was constrained by concerns about the impact of regulation on industrial competitiveness.

As discussed in chapter 1, the intensified focus in EU policy making on the regulatory burden placed on industry has been driven by developments in the realm of institutions, discourse, and politics. The creation in June 2002 of the Competitiveness Council comprised of economics and industry ministers of national governments institutionalized an emphasis within the Council of Ministers on reducing the regulatory burden on industry. In the realm of discourse, the 2000–10 Lisbon process to enhance the competitiveness of industry in the EU increasingly has been invoked as the impetus behind a strategy to stimulate growth and job creation rather than to advance the social and environmental components woven—however ambiguously—into the original language of the Lisbon objectives. Finally, organized industry interests have become more effective and developed additional points of access at the EU level, countering the lobbying impact of environmental interests organized at the European level. All three of these dynamics are fully evident in the case of chemicals regulation.

The Competitiveness Council—formed through a merger of three existing Council configurations (internal market, industry, and research)—emerged from the EU's June 2002 Seville summit of heads of government as an institutional innovation to elevate the focus on industrial competitiveness.[25] REACH was the first legislative proposal placed on the agenda of the Competitiveness Council.[26] Prior to this point, the process of drafting REACH had been conducted by the Council of Environment Ministers. In this context, two factors had operated to place environmental concerns at the core of the proposed regulatory framework. First, environment ministers of several high-standard states served as the driving force behind the proposal. Second, within those governments with large chemicals sectors (Germany first and foremost, but also France, the United Kingdom, and Italy), environment ministers were not primary guardians of industry concerns, in contrast with the economics and industry ministers that came to be represented in the Competitiveness Council.

In its first formal meeting to address the REACH proposal, the Competitiveness Council established that its objective was to ensure that efforts to protect public health and the environment did not impose excessive costs on chemicals producers.[27] Competitiveness ministers invoked the Lisbon agenda and their aim to promote "appropriate regulation which stimulates economic activity and does not hamper it."[28] Over time, the Competitiveness Council came to focus increasingly on the need to take account of industry concerns, to use impact assessment to reduce the regulatory burden imposed by REACH, and especially to examine and mitigate the impact of the REACH legislation on small and medium enterprises (SMEs), where industry interests identified the greatest threat to competitiveness and jobs.

## INDUSTRY LOBBYING AND THE EUROPEAN COMMISSION

Elements of the business sector—particularly well organized and demonstrative in the chemicals sector—responded to the ascent of the precautionary principle with a call to counterbalance precaution through "rational risk management" (Löfstedt, 2004). The industry offensive in response to REACH began in 2002, when the vice-chairman of Germany's chemicals giant BASF, Eggert Voscherau, became president of *Conseil Européen des Fédérations de l'Industrie Chimique* (CEFIC), the European Chemical Industry Council.[29] Three tactics were at the core of the efforts of the chemicals industry federation to weaken REACH, which intensified as the European Commission prepared its formal legislative proposal in 2003. First, CEFIC representatives sought to consistently invoke the commitment made by member state governments in 2000 at Lisbon to dramatically lift the competitiveness of European industry by the end of the decade, insisting that REACH was fundamentally at odds with the Lisbon objective. Second, CEFIC, although led by Europe's largest chemicals producers, deployed to its advantage the EU's strategy of job creation through explicit support for SMEs. CEFIC emphasized the heavy burden regulatory costs would impose on SMEs in the sector, with attendant consequences for business failures and job losses. As discussed below, this focus was later to find resonance in the Competitiveness Council once it took primary responsibility for representing the response of national governments to REACH. In addition to the leverage gained from invoking collective member state commitment, this tactic had the advantage of taking the focus off of large multinational companies for whom it was difficult to make a credible case that they could not absorb the regulatory costs implied by REACH.

Third, CEFIC achieved a dramatic expansion and diffusion of industry interests arrayed against REACH by drawing in the support of firms using chemical substances as inputs (i.e., "downstream users") rather than limiting their forces to chemicals producers themselves. This step was facilitated by the ascent in 2003 of Jürgen Strube, CEO of Germany's BASF, to the presidency of UNICE,[30] Europe's Union of Industrial and Employers' Confederations (Corporate Europe Observatory, 2005). Furthermore, the constellation of downstream users of chemical substances was broad and diverse, including the confederation of national household appliance industry federations (Conseil européen de la construction d'appareils domestiques [CECED]); the federation representing Europe's digital technology industry (particularly communications and consumer electronics) (European Information & Communications Technology Industry Association [EICTA]); and the European Semiconductor Industry Association (ESIA). As one example of the additional leverage this coalition of chemical sector allies gave to the chemicals industry in its effort to ease the rigor of the REACH framework, the federations involved openly opposed an amendment from the EP's Environment Committee that sought to reintroduce mandatory substitution for hazardous substances back into the regulatory framework. Downstream users, drawing on their credibility in evaluating the economic impact of mandatory substitution across a range of industries, couched their opposition in terms of competitiveness, arguing that "[e]nshrining mandatory substitution plans as a pre-condition for authorisation of substances under REACH increases legal and operational uncertainty and thereby risks undermining investment and competitiveness in the EU."[31]

The chemical industry's organized opposition to the REACH proposal intensified pressure on the European Commission to advance its "better regulation" agenda. Chemicals industry advocates commissioned their own studies of the potential impact of REACH, including German and French industry confederation reports that warned of the potential for massive job losses (Corporate Europe Observatory, 2005). Chemical industry representatives and allies called upon the Commission to redraft the regulatory framework provided by REACH in accordance with the Lisbon competitiveness objective. In response, the European Commission initiated a series of stakeholder meetings designed to identify means to reduce the regulatory burden imposed on industry by REACH. In addition, the Commission authorized a series of impact assessment studies by private consultants.

Environmental NGOs (WWF and the European Environmental Bureau) withdrew their support from the core impact assessment, arguing that the approach of the study was biased toward business costs to the exclusion of benefits from innovation and improved worker safety.[32] They also objected to the elimination of the requirement for mandatory substitution of substances of "very high concern" and replacement of that stipulation with a provision that producers demonstrate "adequate control" of such chemicals.

In November 2003, the Commission undertook an "Extended Impact Assessment" and organized an associated stakeholder workshop, and in January 2004 an Ad hoc Working Group on Chemicals—significantly under the auspices of the newly created Competitiveness Council of economics and industry ministers of national governments—began its examination of the REACH proposal. In March 2004, the European Commission under intensifying pressure from industry interests—led by Germany's giant BASF—entered into a "Memorandum of Understanding" with the federation of EU employers (UNICE) and the chemicals industry federation (CEFIC), as well as representatives of the EP, Council of Ministers, and environment NGOs. The MOU established a working group to oversee the commissioning of business case studies that would consider the potential for the REACH regulatory burden to induce withdrawal from the market of some chemical substances. The resulting studies carried out by private consultants indeed found potential vulnerabilities of a substantial number of substances, raising significantly the projected costs and threat to innovation and to SMEs posed by REACH.[33] Reflecting these studies, the Dutch Council presidency presented an overview of 36 impact assessments in October 2004, which warned of the excessive costs of the REACH regulation.[34] NGOs heavily criticized the studies emerging from the MOU for focusing exclusively on risks to business to the exclusion of potential benefits (ChemSec, 2004).

Finally, the Commission carried out a series of "appraisal exercises" narrowly focused on alternative mechanisms of achieving specific regulatory objectives, such as an assessment of alternative approaches to registration of low-volume substances prepared by the Commission services on behalf of the Council Ad hoc Working Party on Chemicals (November 2005). Additionally, RPA consultants, the private consulting firm that carried out the initial impact assessment, conducted several appraisal exercises in 2006 under a REACH Technical Assistance contract with the Commission's DG Enterprise and Industry.

Unsurprisingly, multiple rounds and varieties of impact assessment focused on different elements of the REACH proposal translated into far-reaching revisions to the draft regulation. The initial independent impact assessment of the costs of implementing REACH produced an estimated cost of 12.6 billion euros over the course of the regulatory framework's 11-year implementation period. In response to the impact assessment process, the Commission redrafted its proposal to eliminate more than 10 billion euros of these costs through sharp reductions in requirements for Chemical Safety Reports and information on low-production-volume substances (Extended Impact Assessment: 12).

## Organized Industry Interests and the European Parliament

As the impact assessment process led the European Commission to whittle down regulatory rigor in response to industry demands for a reduced regulatory burden and a discourse that had shifted heavily in favor of lightening the regulatory load, organized chemical industry interests expanded their lobbying efforts within the EP. In particular, the European Chemicals Industry Council, CEFIC, developed a critical point of access within the Parliament that allowed it to counter the influence wielded by organized environmental interests in the initial stages of the REACH process through the EP's Environment Committee.

The response of the EP to the European Commission's draft legislative proposal reflected an emerging internal tension between party groups and between standing committees, and the extent to which environmental NGOs operate in a policy network with the EP's Environment Committee. While ENVI, whose Green Party rapporteur, Inger Shörling, took the lead in drafting the EP report, adopted a position very close to that presented by environmental NGOs like Friends of the Earth Europe and the European Environmental Bureau (EEB), the Internal Market Committee (IMCO), and Industry, External Trade, Research and Energy Committee (ITRE) introduced critical concerns of the chemical industry. The report drafted by ENVI called for the extension of registration to imports in preparations or products below 1 ton (by 2012) and basic registration for all chemicals, regardless of volume, in order to eliminate incentives for manufacturers to avoid registration by remaining below the mandatory threshold (European Parliament, 2001: 25). The draft suggested that data requirements for the registration phase be based on aggregate volume. The ENVI draft also proposed that there be substitution deadlines for substances of very high concern, including those considered carcinogenic, mutagenic, or reprotoxic (so-called CMRs) and those deemed persistent, bioaccumulative, and toxic (PBTs), and, where there is evidence of disruptive properties, endocrine disruptors as well. Similarly, the report suggested a terminal date (11 years) for authorization of use of substances of very high concern; in instances of closed chemical processes with no environmental release, this could be extended to 2020, at which point "discharges, emissions and losses of hazardous substances...shall cease" (European Parliament, 2001: 24).

In contrast with the position of the EP's Environment Committee, the Internal Market Committee[35] urged that the focus of chemicals regulation fall on "problematic uses of a given chemical" rather than intrinsic properties that might be harmful. The committee also asserted that production of chemicals in amounts below 1 ton should be excluded from the registration system (European Parliament 2001: 28). A third committee, ITRE, called for authorization to be limited to CMR and POP (persistent organic pollutants) substances, and explicitly invoked the issue of competitiveness, citing the Lisbon objectives and calling for a balance between protection of health and environment and the need to foster innovation and enhance competitiveness (European Parliament, 2001: 32).

The initial battle over the REACH legislation in the Parliament as its first reading began in February 2005 was a struggle for jurisdiction. The internal rules of the EP establish the relationship between lead committees and those assigned to provide opinions; the responsible committee considers these opinions, and may draw on them in order to craft a position likely to win approval in plenary, but is not required to incorporate them into its report. A specific procedure for cooperation between EP committees (the "enhanced Hughes" procedure, introduced in 2002) was crafted with the expectation that committees would engage in "enhanced cooperation" when a piece of legislation fell astride the substantive realms of two committees, or contained parts encompassing two distinct substantive policy areas. There are three aspects to the process. First, the two committees jointly decide on the timetable for considering the legislation. Second, the rapporteur from the lead committee and the draftsman from the committee engaged in enhanced cooperation "endeavour to agree on the texts they propose to their committees and on their position regarding amendments."[36] The third dimension is the most complex and ambiguous, for it calls upon the chairman of the responsible committee to accept the amendments of the cooperating committee relating to its field of jurisdiction.[37]

The European Parliament's Conference of Presidents (comprised of the President of the European Parliament and the chairs of the political groups) initially assigned REACH to the Environment Committee. Asserting that the REACH regulation was about industrial policy rather than environmental policy, the Internal Market Committee contested exclusive ENVI responsibility for REACH.[38] In response, the Conference of Presidents invoked enhanced Hughes, taking the unusual step of granting *both* the Internal Market and Industry, Research, and Energy committees enhanced status. As I demonstrate below, the additional authority extended to these committees ultimately made it necessary for ENVI to hew closer to the industry position in order to obtain majority support in the full EP.

IMCO's position overlapped with that of the Competitiveness Council of national economics and industry ministers, although IMCO, with a membership broadly representative of the full EP, retained a preference for slightly stronger environmental protection. In order to salvage more rigorous environmental protection without losing its majority in plenary, the ENVI rapporteur, Italian Socialist Guido Sacconi, had to substantially incorporate concerns for industrial competitiveness championed by IMCO. This implied a trade-off between acceptance of the IMCO position on chemicals registration and retention of the ENVI position on authorization of chemical substances. Without retention of the Environment Committee's position on chemicals authorization, the Sacconi Report (i.e., the EP position drafted in the Environment Committee) would have lost *committee* support; without concessions on registration to the Internal Market Committee, the Sacconi Report would have been defeated *in the plenary*. Reflecting the potential for IMCO to defeat the ENVI position, the compromise on registration supported by IMCO was endorsed in plenary by a vote of 438 to 155, with 41 abstentions, while the authorization package reflecting the

ENVI position passed only by a margin of 324 to 263, with 13 abstentions.[39] The latter result left the EP position on authorization of chemical substances vulnerable to challenge by the Council of Ministers.

The chemicals industry federation (CEFIC) initially aimed its lobbying efforts across institutions as well as EP committees, including ENVI. However, with IMCO and ITRE granted enhanced status, CEFIC was able to focus its activities on members of the European Parliament (MEPs) with whom it already had strong relationships. This dynamic is consistent with the findings of research on lobbying congressional committees in the US context, which suggests that organized interests concentrate their lobbying resources on efforts to provide allies with information and arguments to fend off hostile amendments rather than attempting to win over opponents and engaging in "counteractive lobbying" simply to refute the arguments of opponents (inter alia, Hojnacki and Kimball, 1998: 785). As a CEFIC official responsible for EU Government Affairs explained, "During the process it became rapidly clear that in terms of getting our political messages across, we had to work more intensively with the Industry and Internal Market Committees, though you have to ensure that you keep all options open to get the winning compromise."[40]

Environmental NGOs, in contrast, found their network links with Environment Committee MEPs substantially neutralized by ENVI's need to compromise with other committees, specifically IMCO and ITRE. Environmental interest associations did not possess the resources to build anew relations with members of other committees. Moreover, environmentalists experienced a critical asymmetry: while industry federations CEFIC and UNICE enjoyed access to IMCO and ENVI alike, IMCO members expected environmental interests to present their arguments predominantly through ENVI.[41]

The fact that a member of the European Peoples' Party (EPP) from the country with the largest delegation in the EP and the most substantial stake in chemicals sector regulation—Hartmut

Nassauer of Germany—served as IMCO rapporteur for REACH, intensified the pressure on ENVI to yield to IMCO, which had sought full responsibility for the registration portion of REACH under the enhanced Hughes procedure.[42] The pressure on ENVI chair Karl-Heinz Florenz, a German member of the center-right, competitiveness-centric EPP, was particularly acute.

The Environment Committee voted on its own amendments to the Commission proposal, and then did the same for the amendments proposed by all other committees, including IMCO and ITRE. ENVI incorporated a significant share of IMCO amendments into its report. ITRE proposed 287 amendments in its opinion on the Commission's REACH proposal; ENVI incorporated 29 of these into its report. A much larger share of IMCO's amendments—89 out of 335—were included in the Sacconi Report.[43] Illustrating the difficult compromises forced on ENVI, the Sacconi Report was approved in the Environment Committee only by a margin of 40 to 19, with 2 abstentions. Moreover, IMCO used the enhanced status it garnered from the Hughes procedure to induce ENVI to bring the IMCO/ITRE-favored approach to the registration of chemicals directly to the plenary. This step was the product of a deal between REACH rapporteur Sacconi, ENVI chairman Florenz, and the chairs of IMCO and ITRE; the Green Party coordinator in ENVI opposed this solution.[44] Indicating how the enhanced status of the Internal Market and Industry Committees altered the EP's position on REACH, a CEFIC official concludes that "[u]ndoubtedly, had the ENVI Committee been the only responsible committee, it would have been much more difficult to get industry's positions taken into consideration."[45]

In the full EP, an amendment of the European People's Party rejecting the inclusion of substances produced in volumes of less than 1 ton in the registration system narrowly carried the day, as did an amendment reducing the scope of substances to be phased out to those included in the Commission's White Paper—those scientifically proven to be CMRs and persistent organic pollutants.[46] The amended resolution approving the Commission approach with modifications was approved by the full EP by the relatively narrow margin of 242 to 169, with 35 abstentions (Shörling, 2004: 74).

In addition to the gains made by producer interests on the issues of chemicals registration and phase out of hazardous substances, weak plenary support for the Environment Committee's approach to chemicals authorization and substitution left the EP position vulnerable to the demands of the Competitiveness Council. The ENVI position on authorization of chemicals did not survive the EP's second reading, and the final legislation provided for compulsory research and development *plans* rather than compulsory substitution of safer for the most hazardous chemicals. National government ministers operating in the Competitiveness Council had formally begun discussing in fall 2004 the idea of reducing data requirements for chemical substances produced in low volumes in order to mitigate the regulatory cost burden and its impact on industrial competitiveness.[47] In December 2005, competitiveness ministers reached a political agreement on the Council position on the REACH legislation, in which the Council eased requirements for authorization and substitution standards for dangerous substances relative to the original European Commission proposal and the position of the EP.[48]

As the evidence in this section indicates, the emergence of the Competitiveness Council marked an institutionalization in the EU policy-making apparatus of concerns about the impact of regulatory costs on industrial competitiveness that coincided with the debate over REACH. In addition, organized industry interests over time cultivated increasingly effective points of institutional access inside the European Parliament (Smith, 2008).

## Precaution, Cost-Benefit Analysis, and Prospects for Reform of TSCA

Comparative studies of regulation typically have emphasized that the central role of business interests and economic costs in decision making distinguish regulation in the United States from the European regulatory context (Krämer, 2004: 68–69). In their comparative study of European and US chemicals regulation of more than two decades ago, Brickman et al. could write that, as compared with the heavy reliance on economic analysis in US regulation, "[t]he virtual neglect of formal cost-benefit analysis in European decision making presents a striking contrast" (Brickman, Jasanoff, and Ilgen, 1985: 40). However, since the Lisbon summit of 2000, which placed renewed emphasis on external competitiveness of EU industry and an attendant commitment to less burdensome regulation, regulatory impact assessment has gained significance in EU policy making.[49] As Jonathan Wiener points out in his comparison of EU and US regulatory systems, both "have now adopted risk assessment and cost-benefit analysis as basic criteria for new regulations" (Wiener, 2004: 86). And, as we have seen, cost-benefit analysis was at the center of the debate over REACH within the EU institutions.

Furthermore, there is ample evidence that, despite the ineffective nature of the prevailing chemicals regulatory regime, the precautionary principle is hardly foreign to the regulatory debate in the United States. In fact, actors ranging from environmental groups to state legislators to members of the US Congress have attempted to move chemicals regulation in the direction of precautionary policy. In the face of institutional fragmentation of the legislative process and the existence of a large number of points at which more robust regulation might be vetoed, actors seeking to advance a precautionary approach to chemicals regulation have pursued efforts at the state level, judging that these have a higher probability of success than pursuit of new regulatory departures at the federal level. This approach is reflected in the Louisville Charter initiative, which seeks to alter the regulatory landscape by implanting some of the principles at the core of the EU's REACH legislation in individual US states.

While REACH was not the defining source of the Louisville Charter, it has informed the Charter's core principles.[50] The Charter, initiated in 2004, seeks to enshrine in US chemicals regulation the principle of "no data, no market," designed to remedy the paucity of information available concerning the approximately 80,000 substances in the EPA's commercial chemicals inventory, including nearly 3,000 produced at volumes exceeding 1 million pounds annually (Guth, Denison, and Sass, 2005). Additional themes include the phasing out of substances that are persistent, accumulate in the human body, and are highly toxic; and the availability of information concerning chemicals exposure and risk to workers and consumers.[51]

While the Charter establishes desired objectives of chemicals regulation, it does not outline a comprehensive strategy for achieving these goals. The numerous environmental groups attached to the Charter seek to alter the constituency for environmental regulation; the aim is to change federal law indirectly rather than directly. This is to be accomplished through shifts in the constituency for chemicals regulation at the levels of industry and state political representation. Efforts to achieve this shift within industry focus on the purchasing habits of businesses that are large downstream users of chemical products. A primary example is Health Care Without Harm, a campaign initiated in 1996 by a group of environmental and health care associations following the EPA's identification of medical waste incinerators as the leading source of dioxin, a known carcinogen.[52] The campaign's objective is to alter the purchasing patterns of the health care sector, one of the largest users of chemicals products, by promoting substitution for harmful products such as polyvinyl chloride (PVC) plastic in i.v. bags. Shifting product demand of such large downstream sectors as health care can produce market effects that spill over into other sectors.

State-level efforts can set a higher standard for chemicals regulation and also create additional supporters of more rigorous environmental regulation through regulatory measures in individual states. Prominent state-level projects include California SB 484, the Safe Cosmetics Act, signed into law in October 2005. Like Health Care Without Harm, SB 484 emerged from collaboration between health care and environmental interest associations, including Breast Cancer Action, the Breast Cancer Fund, and the National Environmental Trust. While SB 484 does not give the California Department of Health Services sweeping regulatory powers, it does require that cosmetics manufacturers provide information about any ingredients that are on federal or state lists of carcinogenic or mutagenic substances.

In addition to the efforts of environmental interest associations and state-level policy makers to strengthen regulation of chemicals, there have been modest efforts at the federal level that further demonstrate the significant support for a precautionary regulatory approach in the United States. Associated with the Louisville Charter initiative, Senator Frank Lautenberg and Representative Henry Waxman introduced in November 2005 The Child, Worker & Consumer Safe Chemicals Act. The act proposed fundamental departures from existing US chemicals regulation borrowing from the principles of REACH. These include a comprehensive registry of information on chemicals in production, compiled according to a priority list based on production volumes and concerns about health threats, a system of biomonitoring in order to identify dangerous substances, and incentives for the promotion of safer substitutes and green chemistry.[53] The bill was referred to the House Energy and Commerce Committee but died in the Subcommittee on Environment and Hazardous Materials at the end of the 109th Congress.[54]

The proposed legislation called for companies to submit to the EPA within one year chemical safety certification for all substances, and to provide data on the physical, chemical, and toxicological properties of each chemical substance, as well as data on production volume, known uses, and exposure. The act charges the EPA Administrator with development within 18 months of a priority list of 300 chemicals for safety certification. If certification is not completed within five years, the chemical in question may no longer be marketed. The proposal revises the safety standard for chemicals from the "reasonable expectation of unreasonable risk" stipulated in TSCA to "a reasonable certainty that no harm will be caused by aggregate exposure of a fetus, infant, child, worker." (H. R. 4308, Section 503).

The biomonitoring clause requires periodic testing for presence in human blood and tissue for chemicals produced in volumes greater than 1 million pounds annually or for which there is concern about persistence and bioaccumulation. The proposal directs the EPA Administrator to establish incentives for the promotion of safer substitutes for existing chemicals, including expedited review for new substances where the manufacturer documents that the chemical is a safer alternative for a particular use and special EPA designation to aid in the marketing of the new substance (H. R. 4308, Section 506). In addition, the act calls upon the EPA Administrator to set up a network of chemistry and technology research and clearinghouse centers "to support the development and adoption of safer alternatives to chemical substances." Finally, the act contains a strong anti-preemption clause, which protects the rights of states to adopt more rigorous regulations (H. R. 4308, Section 511).

Although H. R. 4308 died in 2005, Congress in early 2009 began hearings on revising TSCA, and the Safe Chemicals proposal was reintroduced in both houses of Congress in 2010. Hearings on TSCA revision began only after the Government Accountability Office added the EPA's protection of public health and the environment from the threat of toxic chemicals to its list of "high risk" government programs.[55] In doing so, GAO warned that "the EPA's inadequate progress in assessing toxic chemicals significantly limits the agency's ability to fulfill its mission of protecting human health and the environment."[56]

The GAO cites the evidentiary standard for EPA regulatory enforcement established by the courts as a primary obstacle to effective chemicals regulation: "[T]he statutory requirements EPA must meet . . . present a legal threshold that has . . . discouraged agency action" (2009a: 2). Accordingly, GAO calls for Congress to enact amendments to TSCA that reduce the evidentiary burden on the EPA, specifically invoking REACH as a model (GAO, 2009a: 6). These include altering the "unreasonable risk" standard for EPA regulation of existing chemicals as well as the requirement that the EPA undertake a cost-benefit analysis of every regulatory remedy and impose the least burdensome regulatory measure (2009a: 12; 2009b: 10).

The version of the Safe Chemicals Act reintroduced in April 2010 focuses on shifting responsibility to manufacturers for testing new chemicals before bringing them to market and submitting health and safety data to the EPA for chemicals currently in use.[57] The debate over TSCA reform in 2010–11 takes place in an altered regulatory landscape in which chemical producers face a variety of restrictive measures from state governments, as well as steps by state governments to use the market power of state agencies to encourage the production of safer products by avoiding certain chemicals in their purchasing decisions.[58]

In response, rather than opposing federal regulatory reform outright, the industry association representing major producers has publicly endorsed a set of principles for TSCA reform that seek to minimize the regulatory burden, including "risk-based" management of chemical hazards (as opposed to precaution); keeping the ultimate burden for determining hazards on the EPA; minimizing additional testing requirements imposed on industry by making use of available data; retention of a cost-benefit basis for EPA regulatory steps; and prioritization and provision of legal certainty to manufacturers in order to safeguard efforts to innovate.[59] The industry approach remains entirely consistent with what has worked in the past—preemption of significant increases in regulatory rigor through support for voluntary measures and minimalist regulatory reform.

CHAPTER 3

# Health Claims on Food Labels: Protecting Free Commercial Speech versus Completing the Single Market

THE CASE OF NUTRITION LABELING CONTRADICTS THE NOTION that regulation in the United States is shaped by a high tolerance for risk. In fact, the modern history of food health claims regulation in the United States reflects a cautious and highly restrictive approach. This approach only began to unravel in the opening years of the twenty-first century, and regulation has become progressively more permissive since. In addition, this chapter echoes the findings of chapter 2 that, in the European Union (EU), producer interests have intensified organization and influence within institutional channels that traditionally have given a receptive hearing to the claims of environmental, health, and consumer protection NGOs. But the chapter also shows that rulings of the European Court of Justice (ECJ) have had very different consequences for regulation of health claims on food labels than court decisions in the United States. This is so even though rulings on critical principles of free commercial speech, protecting competition, and the proportionality of regulatory remedies in the United States and the EU have coincided substantively.

Taking an expansive view of free commercial speech doctrine, rulings in the United States have served to reduce restrictions on health claims. In contrast, ECJ findings that national restrictions on health claims on food labels are inconsistent with EU law—on grounds that they violate the principle of the free movement of goods across borders or are disproportionate to the health objectives sought—have created an impetus for EU institutions to seek European-level regulation and for national health ministers to endeavor to recapture at the European level some of the regulatory control lost to ECJ decisions. Furthermore, producers seeking legal certainty across the single European market have strongly favored regulatory harmonization at the European level—even if seeking less restrictive measures than those supported by EU institutions and national health ministers.

In the United States, the 1990 Nutrition Labeling and Education Act required that packaged foods carry nutrition facts labels and established stringent controls over health claims on labels. The NLEA restricted claims very narrowly to those supported by a body of scientific evidence and linking particular nutrients to prevention of specific disease. Initially, the Food and Drug Administration (FDA) approved only ten substance-disease relations. Yet more than a decade later, just as the EU was moving toward stricter regulation of claims in order to protect consumers from misleading advertising on food labels, the United States was moving toward a far more relaxed regulatory posture.

This trajectory culminated in the 2009 "Smart Choices" program, an industry-initiated program of voluntary regulation that sought to condense and simplify health information for consumers by awarding a green check mark featured prominently on the front of the packaging label of foods meeting certain positive standards for nutrient content and not exceeding specified levels for fat, sugars, and sodium. Almost immediately after its inception, the program was suspended by food producers and distributors themselves following an outcry from consumer and health advocates, a cautionary October 2009 letter to industry from the FDA,[1] and the threat of growing action to protect consumers by state attorneys general in response to the appearance of the program's check mark of approval on sugary breakfast cereals Fruit Loops and Cocoa Krispies, among other highly processed, sugar-, salt-, and fat-laden items widely considered unhealthful by nutritionists.

In contrast with the United States, the regulatory trajectory in the EU in many ways resembled that followed in the chemicals sector—movement in the mid-2000s toward a more rigorous regulatory regime (in this case, one based on nutrition profiles restricting health claims on food labels—an approach followed in the United States until as little as a decade ago under the Nutrition Labeling and Education Act)—but a regime nonetheless constrained by rising concerns expressed in industry lobbying and especially evident in debate within the European Parliament (EP) about the regulatory burden and its impact on the competitiveness of industry.

## The United States and the Transformative Effect of Free Commercial Speech Doctrine

Until the mid-1980s, health claims on foods were severely restricted by the US government. The Federal Food and Drugs Act of 1906, the Federal Food, Drug and Cosmetic Act (FDCA) of 1938, and subsequent amendments, prohibited companies from making false or misleading claims on product labels (Blim, 1994: 737). For nearly eight decades, health claims on foods were treated as implied drug claims; any health claim on a food label would result in the product being treated as a new drug, subject to FDA approval (Pappas, 2002: 27). This treatment of health claims was a deterrent to any efforts to make such claims,[2] amounting to a prohibition of claims.

The environment for health claims on food labels shifted in the mid-1980s, when the Kellogg Company began to employ as a marketing tool for its All-Bran cereal a National Cancer Institute finding linking fiber consumption with reduced cancer risk. While the use of the claim might have led the FDA to pursue regulatory enforcement action, the FDA refrained from doing so due to the National Cancer Institute's backing of the claim (Blim, 1994: 738; Michaels, 1995: 320; Steinborn and Todd, 1999: 404; Pappas, 2002: 27). The absence of FDA action signaled a more permissive environment for health claims on food labels, and such claims began to proliferate in the late 1980s (Steinborn and Todd, 1999: 405). Widespread criticism of the FDA ensued, with consumer associations and health organizations lobbying Congress for legislative action to restrict the use of health claims. The ultimate result was the Nutrition Labeling and Education Act of 1990 (NLEA).

Organizations representing food industry interests also favored federal action that led to the NLEA. The multiplication of health claims in the wake of the Kellogg All-Bran decision led to action by some state governments, which, while possessing no power to enforce the 1938 FDCA, could nonetheless enact state laws to regulate food labeling in the interest of protecting citizen welfare (Bradley, 1994: 652–653). State consumer protection activity created an environment of regulatory diversity and legal uncertainty for food producers, who sought greater regulatory uniformity and legal certainty (Bradley, 1994: 653). Meeting this aspiration of businesses in the food sector, the NLEA contained a clause preempting states from enacting more stringent regulations, though it empowered states to enforce federal law in federal courts (Bradley, 1994: 657–658; 661–662).[3] Given the paucity of resources available to the FDA, it in fact depends on states to enforce the provisions of the NLEA (Bradley, 1994: 660; 662).

The NLEA, which took effect in 1993, authorized the FDA to approve health claims linking consumption of specific substances to improvements in health or disease prevention. The wording of the Act brought a very broad range of health- and disease-related claims under its terms (Steinborn and Todd, 1999: 407). No longer would all foods with labels containing health claims be treated as new drugs. However, in order to obtain FDA approval, claims would have to be supported by "significant scientific agreement." Congress extended to the FDA discretion to determine how this standard would be interpreted. In response, the FDA adhered to a strict interpretation of the "significant scientific agreement" criterion, allowing only a small number of health claims. The FDA approved seven claims under the NLEA in 1993,[4] and only five more from 1993 to 2001. The NLEA allowed the FDA to deny claims supported by some scientific studies if it determined that there was not significant agreement among experts concerning the claimed benefit, and to disallow even those claims meeting the standard for scientific evidence when the substance in question contained a "disqualifying" level of another nutrient (excessively high for some nutrients, such as sugar and sodium, but too little of others) unrelated to the health claim (Blim, 1994: 735, 740).

Critics of the FDA's restrictive approach—largely food-trade associations such as the National Food Processors Association (NFPA), the Grocery Manufacturers Association, and the American Bakers Association—charged that the FDA was deterring claims applications by taking far longer to assess claims than mandated by the NLEA, requiring exhaustive documentation of studies relating to the health claim in question, and rewriting wording for submitted claims that rendered claims wordy, cumbersome, and of little marketing value (Pappas, 2002: 28). Seeking to streamline the FDA approval process, Congress in 1997 passed the Food and Drug Administration Modernization Act (FDAMA). The Act called for the FDA to authorize those claims backed by the authoritative statement of another government agency, such as the National Institutes of Health (Pappas, 2002: 29).

Passage of the FDAMA notwithstanding, it was the decisions of courts rather than congressional action that eroded the strict regulatory regime established by the NLEA. In the 1999 *Pearson v. Shalala* case, marketers of dietary supplements challenged a ruling by the FDA that none of their proposed claims met the "significant scientific agreement" standard.[5] The litigants had proposed to the FDA that rather than rejecting the proposed health claims outright, FDA should require a disclaimer expressing its view that the evidence supporting the claims was inconclusive. After a district court ruled against the supplement marketers, they appealed to the US Court of Appeals for the District of Columbia. In its reversal of the district court, the appellate court judged that while the FDA indeed had an interest in protecting consumers against misleading claims through regulatory action, there was not a reasonable fit between the government's interest and the means it chose to achieve its objective. In particular, the court rejected the government's argument that the First Amendment protection of commercial speech does not contain a preference for disclosure over suppression. Based on its grounds for rejecting the proposed health claims, the FDA could have attained its objective of consumer protection through less restrictive means—the use of a disclaimer rather than outright suppression of speech (Pappas, 2002: 30).[6] As discussed later in the chapter, this reasoning paralleled very closely the logic of ECJ decisions taking place at around the same time, which struck down national laws regulating health claims on the grounds that they were excessively restrictive and disproportionate to the objectives they sought.

As David C. Vladeck argues (1999: 541), while the protection of commercial speech was by the time of the *Pearson* decision well established, the court's decision took a newly restrictive approach to the ability of the government to advance critical objectives through limits on commercial speech. Still, since the *Pearson* case involved dietary supplements rather than conventional foods, the FDA persisted in its view that it could pursue distinct regulatory regimes for the two categories (Pappas, 2002: 29). Previously, the FDA had imposed an even more restrictive regime for dietary supplements; after *Pearson*, the FDA proposed to sustain its regulatory approach to health claims on foods while acknowledging the implications of *Pearson* for regulating supplements, creating a less restrictive regime for the latter (Pappas, 2002). However, *Pearson*, along with subsequent rulings confirming Pearson's logic, opened the door to a general assault on the FDA's strict regulation of health claims on the grounds of infringement of commercial speech (Pappas, 2002: 31).

In the December 2002 *Whitaker v. Thompson I* decision, the US District Court for the District of Columbia once again ruled against the government and instructed the FDA to approve a claim for a nutritional supplement with an appropriate disclaimer when it attempted to prohibit a health claim involving antioxidant vitamins and cancer on the grounds that the scientific evidence against the claim was weightier than the evidence favoring the claim, rendering the claim misleading.[7] While acknowledging its need to defer to the technical assessment of the FDA, the court judged that the FDA had discounted some studies, implying that it had not undertaken the required overall review of the available evidence, and that the FDA had not provided evidence that a disclaimer—as opposed to outright suppression of the claim—would be confusing to consumers.[8] Accordingly, the FDA's rejection of the health claim was unnecessarily restrictive, and the FDA was instructed by the court to develop appropriate disclaimers for the claims or to demonstrate that *no* disclaimer would correct the misleading nature of the claim.[9] *Whitaker* therefore implied that the FDA would have to provide empirical evidence of the misleading nature of a health claim even where the evidence supporting the claim was weaker than the opposing evidence, leaving a very narrow window for the outright prohibition of a claim and placing a high evidentiary burden on the FDA.[10]

In recognition of the legal tide flowing against the existing regulatory regime, the FDA in December 2002 announced that it would apply the *Pearson* decision to conventional foods. The FDA then established a task force on Consumer Health Information for Better Nutrition, to focus on crafting an approach to regulating qualified health claims on food labels, in contrast with the "unqualified claims," permitted for nutrient-disease relationships meeting the "significant scientific agreement" standard established in the 1990 NLEA. The FDA's proposed approach was to allow claims with some scientific support, subject to three different levels of qualifying language for claims falling short of the NLEA's "significant scientific agreement" measure: scientific evidence that is not conclusive; limited scientific evidence; and little scientific evidence. In the ensuing comment period, health and consumer protection advocates lined up against the idea of qualified claims. Public Citizen and the Center for Science in the Public Interest argued that the FDA's proposed approach to qualified claims "undermines the protections afforded to consumers by encouraging companies to seek permission to make health claims based on preliminary evidence, as opposed to waiting until the evidence demonstrates the existence of significant scientific agreement."[11] Additionally, the congressional authors of the NLEA criticized the FDA's action as an unwarranted departure from legislative requirements and "an invitation for misleading claims on foods."[12]

Conversely, food industry interests endorsed the concept of qualified claims and called upon the FDA to streamline the approval procedure. In a joint comment, the Grocery Manufacturers of America, the Snack Foods Association, the Institute of Shortening & Edible Oils, Inc., and the National Restaurant Association called for the FDA to develop a new regulatory procedure in which applicants would apply directly for approval of a qualified health claim, rather than first having the petition for a claim being denied on grounds of not meeting the "significant scientific agreement" standard and subject to the FDA's enforcement discretion. The FDA, these industry interests argued, would save resources by considering claims based on "credible scientific evidence" independently of petitions for unqualified health claims where the petitioner is aware that the claim does not meet the standard of "significant scientific agreement." Furthermore, industry interests called for a regulatory approach that would incorporate qualifying language into claims rather than imposing "frank disclaimers... that conflict with the claim."[13]

Ultimately, though, it was the decisions of courts that made it difficult for the FDA to resist the concept of qualified claims. The report of the FDA's task force noted that although Pearson specifically involved nutritional supplements, the decision to apply the Pearson logic to foods substantially was motivated by the prospect of additional legal challenges and the expectation that outcomes of any challenges to restrictions on food health claims would be similar. While the FDA wished to provide consumers with better nutrition information, the decision also "was... based on a desire to avoid further litigation over the constitutionality of the health claims provisions of the NLEA applicable to conventional food labeling to the extent that these provisions do not permit qualified claims."[14]

The first qualified health claim was approved by the FDA in July 2003, involving the relationship between the unsaturated fats in tree nuts and a reduced risk of heart disease. Following this decision, the FDA began accepting additional petitions for qualified health claims; those claims found to be supported by some scientific evidence would be subject to very specific disclaimers associated with the level of demonstrated evidence. Two outcomes followed from the FDA's interim approach to regulating disease claims. First, a proliferation of qualified claims emerged over the next several years. Second, the FDA in fact frequently exercised enforcement discretion, either rejecting petitions for claims outright or subjecting claims to detailed qualifying language. In a content analysis of products on the shelves of several large grocery stores, Paula Fitzgerald Bone and Karen Russo France (2009: 256–257) found that manufacturers use qualified health claims far less frequently than legally possible—whether because they are reluctant to introduce consumer consideration of unpleasant health conditions through their labels or because of the limiting nature of required FDA disclaimers and the cost of seeking FDA approval for claims. The large number of qualified claims denied by the FDA included the link between lycopene and prostate cancer; calcium and breast cancer; green tea and a wide variety of cancers; green tea and cardiovascular disease; eggs with omega-3 fatty acids and heart disease; and vitamin E and heart disease.[15]

Even with the opening to qualified claims, then, the FDA attempted both through rejections of petitions for permissible claims and an emphasis on qualifying language to sustain regulatory rigor for claims linking nutrients in foods and disease prevention (Carver, 2008: 151). In other words, the approach of the FDA to qualified health claims on food labels offers additional evidence that the increasingly permissive environment for health claims on food labels in the United States is not simply a product of an aversion to the principle of precaution in the policy-making process.

However, the FDA's exercise of enforcement discretion produced a new round of legal challenges to the restrictive disclaimer language it sought to impose on food manufacturers. Representative of the tenor of these cases, *Alliance for Natural Health U.S. v. Sebelius*, decided in May 2010, again resulted in a D.C. District Court decision that the FDA had acted unlawfully in limiting a series of qualified health claims concerning selenium and reduced incidence of certain cancers proposed by Wellness Lifestyles, Inc. In part, the court held that FDA erred when it "completely eviscerated the plaintiff's claim" rather than adopting a less restrictive approach such as drafting "short, succinct, and accurate disclaimers."[16] Ultimately, court decisions establish that the FDA cannot use disclaimer requirements as a de facto ban on commercial speech that it wishes to suppress; disclaimers, in short, must consist of the minimally restrictive language to accomplish the FDA's objective of protecting consumers from being deceived (Samp, 2003: 328–329). And from the *Whitaker* decision on, the position adopted by industry was that the FDA could deny a petition to make a health claim only where no or little evidence exists in support of the claim; even in the case of very modest credible evidence, the FDA would bear the burden of having to establish that a disclaimer would confuse consumers if it wished to suppress the commercial speech involved.[17]

Still, the FDA remained more or less defiant in upholding regulatory standards under severe legal constraints. In its January 2009 updated guidance for industry on the scientific evaluation of health claims, the FDA emphasized the persistence of the "significant scientific agreement" standard for health claims.[18] In addition, the FDA continued to express its unease with the court's finding in the 1999 *Pearson* decision that a disclaimer attached to the nutrient-disease claim involving antioxidants and cancer was a sufficient remedy to the FDA's assessment that the relationship did not meet the "significant scientific agreement" standard. As the FDA noted in its guidance for industry:

> There is . . . a more fundamental problem with allowing qualified health claims for individual nutrients based on studies of foods containing those nutrients than the problem the D.C. Circuit held could be cured with a disclaimer. Even if the effect of the specific component of the food could be determined with certainty, recent scientific findings on the complex nature of nutrient-food interactions and on the relationships among diet, biological parameters, and disease indicate that nutrients found to have health benefits when consumed in one food or group of foods may not necessarily have the same beneficial effect when they are consumed in dietary supplement form or in other foods.[19]

The history of regulation of health claims on food labels reflects an FDA long committed to a rigorous regime that it simply has been unable to sustain in the face of political and legal challenges. Many legal scholars portray the FDA as rigidly adhering to an untenably restrictive regulatory approach—one that is excessively paternalistic (Fealk-Stickler, 2005: 96), betrays the intentions of Congress, denies valuable information to consumers, and, ultimately, violates the First Amendment protection of free speech. Steven B. Steinborn and Kyra Todd argue that even after the NLEA, "the agency continued to develop policies that largely thwarted efforts to utilize the food label to convey health-related information to consumers" (Steinborn and Todd, 1999: 404). Clement Dimitri Pappas asserts that "[t]he history of FDA's regulations on health claims demonstrates the depths of FDA's institutional resistance to health claims. Congress has told the FDA to change its approach. Courts have told the FDA to change its approach. Yet the agency has done everything in its power to restrict the number and scope of health claims" (Pappas, 2002: 32). The FDA's behavior is not at all consistent with the notion of a cultural proclivity to tolerate risk.

## EU: Producer Lobbying and Recapture of Regulatory Control by National Health Ministers

### National Regulatory Diversity and the Push for Harmonization

Through the 1990s, food labeling regulations in EU member states varied widely, with highly restrictive rules regarding health claims on labels in some countries (Austria, for example) and a more permissive environment in others (Childs, 1998). France and Germany maintained systems in which health claims could be approved on a case-by-case basis.[20] Voluntary codes developed in Sweden and in the United Kingdom were formative for the EU-level regulation that later emerged; the state-funded Netherlands Nutrition Center also crafted a self-regulatory code in cooperation with industry and consumer groups (Hawkes, 2004: 28). Sweden was a leader in health claims regulation, adopting in 1990 a code of conduct established between industry (the Swedish Food Federation and the Swedish Food Retailers Federation) and public authorities, with the Swedish Nutrition Foundation serving as a coordinating and advisory body for the Food Sector Code of Practice. The Swedish Code, revised in 1997, allowed for product-disease claims to be made following two principles: (1) a two-step process, involving a statement of the diet-health relationship, followed by information on the dietary qualities of the product; and (2) a grounding in published scientific evidence. The first edition of the code issued in 1990 allowed for nine generic health claims (Asp and Bryngelsson, 2007: 108).

A voluntary code of practice similar to the Swedish Code emerged through collaboration among industry, consumer groups, and public authorities in the United Kingdom in 2000. The Joint Health Claims Initiative (JHCI) was a response to a growing disjuncture between a regulatory environment in which foods were treated much as they were in the United States under the 1906 Federal Food and Drugs Act, making disease claims for foods illegal, on the one hand, and advances in food science, on the other (Ruffell, 2001). The JHCI established a committee of independent experts to review the evidence for intended claims, and to approve generic versions of claims available to all producers wishing to use the claim. This process would provide significant legal cover to producers following the JHCI Code of Practice, since the JHCI expert committee was to render decisions based on the totality of available scientific evidence, presumably enabling producers to comply with the ban on false or misleading claims contained in Britain's 1990 Food Safety Act (Ruffell, 2003).

Although some member states, such as Austria, maintained stringent restrictions on the use of health claims on food labels, several cases brought before the ECJ produced rulings that struck down the most restrictive regulatory measures on grounds that they interfered with the free movement of goods in Europe's single market and were disproportionate to their consumer protection objective. Projecting from experience in the United States, a more permissive regulatory environment might have ensued in the countries affected by ECJ judgments, followed by a proliferation of health claims. But this did not in fact occur. The critical distinction from the United States is that ECJ decisions took place within the context of the construction of Europe's single market. Court decisions rejecting some national regulations generated a demand from the business sector for greater legal certainty through uniform European rules; this is not entirely distinct from the call from elements of the business sector for uniform national regulation in the United States. But in the EU context, community national health ministers also stepped up support for restrictive regulation at the EU level.

Furthermore, a European Commission committed to advancing the cause of European integration ensured that the demands of the business sector and of health ministers would translate into EU-level regulation. In a 2001 discussion paper focused on nutritional and functional claims (claims identifying the beneficial impact of a particular nutrient, as distinct from health claims, which posit a link between consumption of a food and impact on a specific disease) on food labels, the European Commission's Safety and Consumer Protection Directorate General (DG SANCO) identified the harmful impact of varying rules on the single market:

> In view of the proliferation of the number and type of claims appearing on the labels of foodstuffs and in the absence of specific provisions at European level, some Member States have adopted legislation and other measures to regulate their use. This has resulted in different approaches and in numerous discrepancies both regarding the definition of the terms used and the conditions warranting the use of claims. These discrepancies could act as barriers to guaranteeing a high level of consumer and public health protection, and could constitute obstacles to

the free movement of foodstuffs and the proper functioning of the internal market.[21]

In May 2000, the European Commission brought a complaint to the ECJ asserting that Austria's law on trade in foodstuffs (the 1975 Lebensmittelgesetz, or LMG) violated a 1979 Council of Ministers Directive on food labeling.[22] The directive established packaging information requirements and banned misleading claims about the properties of a food product or its ability to alleviate disease in both packaging and advertising. In order to prevent national requirements for packaging information from becoming a barrier to trade, Article 15 of the directive stipulated that member states could not block trade in foodstuffs conforming to the requirements of the directive, with an exception provided only on public health grounds. The Commission contended that the Austrian law, which banned all health-related claims from food labels without prior authorization from Austria's Federal Minister for Health and the Environment, violated Article 15 as well as the European Economic Community principle of free movement of goods. The Commission argued, in short, that under the Austrian LMG, foodstuffs lawfully manufactured in other member states could not be marketed in Austria (Bulterman, 2003: 65).

In January 2003, the ECJ ruled that the LMG was disproportionate to its stated consumer protection objective because it banned all health claims without prior authorization from the Austrian government.[23] The Austrian government's objective of protecting consumer health, the Court judged, could be achieved with less restrictive measures (Bulterman, 2003: 66). A similar logic prevailed in the ECJ's July 2004 Douwe Egberts ruling. This case was brought by one coffee company against a competitor marketing a product based on claims of weight loss and fat-reduction benefits. The Court found that a 1980 Belgian law on the advertising of foodstuffs, which had the effect of banning the weight-control reference infringed the EU's 2000 Labelling Directive. The 2000 Council Directive outlawed misleading claims; because of the Belgian law's reliance on absolute prohibition of claims regardless of whether or not they were judged misleading, the Belgian law was considered unduly restrictive. As an alternative to outright prohibition, regulatory authorities had the option of employing the less restrictive approach of obligating the producer to supply evidence supporting the claim. The regulatory remedy was accordingly disproportionate to the objective of protecting human health.[24]

Some observers imply that there is a contradiction in the fact that subsequent EU regulation on health claims adopted important elements of the restrictive Austrian law, given that the LMG was ruled disproportionate by the ECJ (Kossdorff, 2006: 526). The apparent contradiction is resolved, though, by the point that European-level regulation simultaneously allows national health ministers to achieve regulatory objectives and facilitates the desire of EU institutions to integrate rather than fragment the European market. In other words, while outcomes such as the LMG and Douwe Egberts rulings have constrained national consumer protection measures as they have in the United States, such cases have served as catalysts for EU institutions to expand the EU's regulatory ambit by sharply delineating areas where the single market remains incomplete. The result in the case of health and nutrition claims was that the European Commission in July 2003 put forward a proposal for a new health claims regulation permitting claims under conditions that would be uniform across EU member states.

## EU Regulation and Producer Mobilization

The European Community first regulated nutrition labeling of foods with a 1990 Council Directive (requiring member state governments to observe the objectives of the legislation, but not stipulating mandatory means of implementation, as for an EU regulation).[25] Designed to prevent the content of nutrition labeling from serving as a barrier to the free flow of goods in Europe's developing single market, the directive established uniform labeling standards. In general, nutrition information provided on packaging would be voluntary, except in cases where a nutrition claim appeared on the label or in advertising of the product. Furthermore, national governments were not to impose labeling requirements beyond those contained in the directive. Nutrition claims were limited to those relating to energy value and a series of nutrients (protein, carbohydrates, fat, fiber, and sodium), as well as vitamins and minerals enumerated in the directive. The directive set out the precise information to be provided in cases of nutrition claims and the form in which claims were to be presented.

In 2000, as the EU revised its 1979 directive on the labeling of food ingredients, origins, and durability, the European Commission began to develop guidelines for a comprehensive food safety regime. Health claims on food labels had begun to proliferate as producers saw an opportunity in rising consumer health consciousness to use claims as a marketing device. In response to increasing demands from national health ministers and industry associations alike, the Commission's DG SANCO called for harmonized health claims rules in order to facilitate free movement of goods in the single market (Commission, 2001: 4). In its 2000 White Paper on food safety, DG SANCO expressed a preference for maintaining the 2000 Labelling Directive's ban on any disease claims on food labels (i.e., claims associating a food product with preventing, treating, or curing a disease), at the same time indicating a willingness to consider allowing functional claims—that is, claims identifying the beneficial impact of a particular nutrient—as well as nutrient claims highlighting the presence of a level of a nutrient in a foodstuff (Commission, 2000: paragraph 101).

As the Commission proceeded with plans to draft legislation on health claims, industry associations mobilized at multiple levels, presenting their positions to the Commission, lobbying the Council, and gaining access to the European Parliament both through political parties and parliamentary committees. In its response to DG SANCO's 2001 discussion paper on nutritional and functional claims, several industry associations expressed a desire for harmonization of rules at the European level. The Confederation of the Food and Drink Industries of the European Union (CIAA) supported the concept of a comprehensive regulatory framework at the European level governing all food label claims, including health claims—but urged that the regime be permissive, with allowable claims proportionate to the level of scientific substantiation available.[26] CIAA, in other words, hoped for the introduction of qualified health claims, which were gaining a foothold in the US regulatory regime at approximately this time.

The Food and Drink Federation (FDF), the UK industry association, endorsed a common regulatory approach coinciding with the structure of Britain's Joint Health Claims Initiative, in which firms sought premarket advice to ensure the substantiation of health claims, but were not required to obtain premarketing authorization. While the FDF advanced the British approach contained in the JHCI, the Swedish Food Federation endorsed adoption at the European level of the Swedish scheme. This more restrictive approach distinguished between generic nutrient-content claims associated with well-established science, which could proceed on the basis of notification to authorities, and product-specific health claims, which would have to be based on testing and premarket approval.[27]

The US National Food Processors Association also contributed a response to the Commission discussion paper, in which it strongly advocated for international harmonization of rules governing health claims. While calling for standards to prevent misleading information, the NFPA, like the CIAA, underscored its preference for "the broad use of nutrient claims."[28] Furthermore, the NFPA urged a process of harmonization in which the Commission would recognize without any additional authorization procedure claims already approved in third countries, such as the United States.[29] The NFPA also opposed adoption of review and authorization procedures for claims in which specific product claims would be approved separately from the general nutrient-health benefit relationship; NFPA, in other words, opposed an authorization system not allowing claims for specific products.[30] NFPA's position represented a response to the European Commission's proposed consideration of the Swedish system of "two-step" assessment and authorization of health claims, requiring a statement about the relationship between a nutrient and a disease outcome, and a second statement concerning the presence of the nutrient in the particular food product.

The European Commission in 2003 proposed a regulation that would establish uniformity throughout the single European market and restrict the use of health claims on food packaging. In its draft, the Commission noted how the proliferation of claims and differentiation of regulatory approaches across member states undermined Europe's single market: "These discrepancies can act as barriers to guaranteeing a high level of consumer and public health protection, and can constitute obstacles to the free movement of foods and the proper functioning of the internal market" (Commission, 2003: paragraph 2).

Two elements were central to the proposed regulation. First, claims about the nutrients contained in a food product or about the link between the product and health could only be made on the basis of generally accepted scientific data, and the latter—so-called disease-reduction claims—would require a premarketing authorization review by the European Food Safety Authority (EFSA).[31]

Second, producers of foods having high levels of fat, sugar, or salt would not be allowed to make claims about beneficial health effects. Such "nutrition profiles" (drawing on elements of Swedish regulation, where health claims are limited to foods meeting certain health criteria) became the most contentious part of the proposal. Reflected in the position of the European

Federation, the CIAA, Britain's Food and Drinks Federation expressed complete rejection of "nutrition profiles." Invoking an argument that would feature prominently in the debate over the piece of legislation ultimately proposed by the European Commission, the FDF indicated that individual foods cannot be classified as "good" or "bad," since many foods with healthful elements also contain ingredients that do not contribute to a healthier diet, and the core objective of regulatory efforts should be overall dietary balance rather than the composition of individual foods in any case.[32] Taking a similar position, the CIAA urged that there simply be a requirement of scientific substantiation for all health claims.[33] The European Consumers' Organisation (BEUC), in contrast, lobbied heavily in support of the notion of nutrition profiles in the legislation, applauding the attempt to reign in the welter of false and misleading health claims that threatened to encourage bad dietary habits among EU consumers (BEUC, 2003).

As the Commission's proposal made its way through the legislative process, organized industry interests succeeded in elevating attention to economic competitiveness in the debate over nutrition claims. As for REACH in the chemicals sector, which was marked by the intersection of environmental/health and economic competitiveness concerns, the European Parliament's Conference of Presidents responded to the joint prominence of health and competitiveness issues by granting the Environment and Internal Market Committees shared responsibility for the nutrition and health labeling legislation. The consequences were similar to those in the case of chemicals regulation—the step broadened industry's access in the EP, with material consequences for the position of the EP on the legislative proposal.

The Environment Committee (ENVI) rapporteur, Adriana Poli Bortone, favored a form of the nutrition profiles contained in Article 4 of the Commission's proposal (European Parliament, 2005: 38–39). However, the Internal Market Committee (IMCO) became a focal point for producer lobbying, and channeled into the debate producers' framing of the nutrition labeling regulation as an instance of excessive regulation, an infringement on consumer choice, and a threat to the competitiveness of the European food industry.

Industry lobbying focused on the impact of labeling restrictions on innovation and competitiveness. Intensified producer lobbying even divided the Parliament's ENVI, often depicted by observers as the primary point of influence in EU policy making for environmental, health, and consumer interests (Collins, Burns, and Warleigh, 1998). Pressure from the German delegation of the Parliament's center-right European People's Party group (European People's Party-European Democrats [EPP-DE]), the EP's largest party group, both through the Internal Market Committee and its members in the Environment Committee, contributed to the division. This dynamic was reflected in the ENVI's April 2005 report to plenary for the first reading of the bill. The report, which dropped Article 4 on the compulsory labeling of nutrient profiles, passed the committee on a 30 to 15 vote. Critics from within the committee attributed the outcome to lobbying by large industrial food companies from Germany and the United Kingdom.[34]

In the EP's first-reading debate, the Industry, Research and Energy Committee's draftsman, German EPP-DE MEP Angelika Niebler, invoked the Lisbon process, focused on improving the competitiveness of EU industry, as a basis for deleting Article 4: "How this proposal for a regulation is supposed to be compatible with the Lisbon strategy, heaven knows."[35] Other members of the European Parliament (MEPs) echoed this concern, warning that the nutrition labeling regulation would undermine innovation in food manufacturing, destroy jobs, and hurt small and medium enterprises (SMEs). While the report of a divided ENVI was adopted in plenary by a margin of 458 votes to 116, with 15 abstentions, the amendment deleting Article 4 garnered 303 votes, with 286 against and 10 abstentions. MEPs from Germany and the United Kingdom led the effort to cut the provision for nutrition profiles from the regulation.[36] In addition, the Parliament voted to move from an authorization procedure to a less restrictive notification requirement for health claims, shifting the burden of proof to the Commission to review claims and decide whether to refer them to the European Food Safety Authority for further assessment. However, the slender margin by which the EP defeated the nutrient-profiles provision in its first reading enhanced the prospect that national health ministers could restore nutrient profiles; such a step by the Council would require an absolute majority of MEPs to reintroduce their nutrition-profiles amendment in the Parliament's second reading.

## National Health Ministers Reclaim Regulatory Control

Within the Council of Health Ministers, the German government initially opposed nutrition profiles. In initial meetings of the Council Secretariat's Working Party on Foodstuffs, the German representation proposed striking Article 4 outright from the regulation. In addition, Germany opposed the authorization procedure for disease-reduction claims, calling for the use of a notification procedure instead.[37]

However, the German government found itself more or less isolated, and its position untenable. As a consequence, the final outcome of the legislative process for the health claims regulation represented a sharp contrast with the outcome in the chemicals sector. In the case of REACH, as shown in chapter 2, EU member state governments concerned with protecting the chemicals sector from burdensome regulation leveraged the weak EP first-reading majority to win additional concessions in the final legislation that departed sharply from the hazardous chemicals authorization provisions the Environment Committee had fought to keep in the regulation. In the instance of nutrition labeling, the Council of EU Health Ministers was in contrast decisive in securing more rigorous regulation, restoring Article 4 on nutrition profiles in their common position of June 2005.

The commitment of national health ministers to restoring the nutrition-profiles provision to the regulation shifted the negotiating dynamic in the second-reading debate within the Parliament and diminished the potency of industry lobbying. In the Parliament's second reading, opponents of nutrition profiles were more or less resigned to modestly amending Article 4 rather than removing it from the bill, while proponents were more assertive than in the bill's first reading. The provision for nutrition profiles remained in the final regulation of December 2006, including the EP's changes to speed the authorization process for claims; to allow claims for some products exceeding the nutrition-profile allowances with an appropriate warning statement; to assist SMEs in adapting to the regulation; and to require Parliament's approval of the nutrition profiles devised by the European Commission.[38] National health ministers ultimately were able to recapture at the European level the regulatory control they had lost through ECJ decisions ruling restrictive national food labeling regulations incompatible with EU market rules on the free movement of goods.

Even in the Parliament's May 2006 second-reading debate, members of the European People's Party (EPP) expressed strong distaste for nutrition profiles on the grounds that they constituted excessive regulation and would have damaging consequences for the industry's competitiveness—a reversion to their initial position that EU laws prohibiting false and misleading advertising made the nutrition and health claims regulation superfluous.[39] Renate Sommer, lead spokesperson for the EPP and a German MEP, suggested that the tendency of the regulation would be "to deprive citizens of their freedom of choice by categorising foods as 'good' or 'bad' on the basis of mythical nutrient profiles."[40] Another German member of the EPP group, Horst Schnellhardt, echoed his opposition to identifying "good" and "bad" foods, asserted that the provision for nutrition profiles amounted to bureaucratic overreach and noted that the EP had no indication of how the EFSA would develop such profiles. Arguing along similar lines, Avril Doyle, head of the Irish delegation to the EPP, criticized "nutrient profiling" for failing to consider consumers' broader consumption habits and suggested that the measure violated the Lisbon competitiveness agenda objectives of "simple, clear and enforceable" regulation. Reflecting the dominant position within the German delegation, Liberal Democrat Holger Krahmer lamented the poor choices confronting the EP because "the Council ignored this House's abundantly clear vote on the health claims at first reading."[41]

But both MEPs opposed to and in favor of nutrition profiles acknowledged during the second-reading debate the critical role of the Council's common position in determining the outcome. German MEP Renate Sommer underscored that the Council's position—"every bit as revolting as the Commission's original draft"—left the EP with little alternative to the compromise assembled by the EP rapporteur. Another German member of the EPP, Thomas Ulmer, noted that "the Council and the Commission have played from . . . a position of excessive strength vis-a-vis Parliament, which has now been forced to give its agreement for fear of even greater damage being done."[42]

Rapporteur Adriana Poli Bortone, citing the Parliament's achievement in gaining modest revision to the nutrition-profiles provision, indicated that the Council's position remained immovable until days prior to the EP's vote: "In essence, the common position stipulated that no type of labelling—neither nutrition nor health labelling—could be used on a product that did not comply with the nutrient profiles established by EFSA."[43] Other MEPs more explicitly cited the reduced impact of industry lobbying in the second reading and credited the Council with shifting the tenor of the debate. Dorette Corbey, a Dutch member of the European Socialists (PSE), the EP's second-largest party group, indicated that "[a]t first reading, Parliament gave in to pressure from businesses that were scared of being exposed." Similarly, Welsh MEP Jill Evans of the Greens noted that at first reading, the EP "voted by a majority to follow the industry line and deleted elements which were absolutely essential to the effectiveness of th(e) legislation." Dutch MEP Kartika Tamara Liotard of the European United Left asserted that "the tenacious lobby of food and particularly the drinks industry has not been as effective this time around," and asked rhetorically whether the EP second-reading vote could be attributed to the fact that "[m]embers are simply embarrassed about the fact that for once, the Council was more progressive than Parliament?"[44] And, in a still more caustic contribution to the debate, Frédérique Ries, a French Liberal Democrat, pointed out on behalf of the EP: "[W]e have nothing to gain from European law being drafted at the International Sweets and Biscuits Fair in Cologne."[45]

## The Persistence of Producer Mobilization

The Regulation on Nutrition and Health Claims on Foods entered into force in January 2007; the European Commission was scheduled to have in place provisions for the nutrition profiles stipulated in the regulation within two years and was supposed to compile a list of health function claims based on generally accepted scientific evidence by January 2010. The Commission was unable to meet either deadline, a consequence of both the vast number of claims submitted and continued producer lobbying to influence nutrition-profiling rules and the range of accepted health claims. By January 1, 2008, member state governments had submitted to the European Commission more than 44,000 proposed claims, which the Commission then consolidated and forwarded to the European Food Safety Authority.[46] Producers objected when EFSA produced its first set of decisions in October 2009 and announced its need for an additional two years to assess all submitted claims, arguing that releasing decisions in batches rather than all at once would create legal uncertainty and distort competition between those whose claims had been rejected and those who could continue to use claims while they were under assessment.[47]

While the Commission initially planned to announce decisions in phases in order to advance the consumer protection objectives of the regulation by removing rejected claims from food labels, it changed course in response to producer objections, announcing in September 2010 that it would publish the list of permitted claims in one go. While the Confederation of Food and Drink Industries of the EU applauded the decision as a boon to fair competition and innovation, consumer interest associations, pointing out that 80 percent of submitted claims had to this point been rejected by EFSA, deplored the delay in advancing consumer protection.[48]

The delay in agreement on a framework for nutrition profiles also provided time for an additional attempt by opponents to strike the nutrition-profiles provision from the regulatory landscape. In spring 2010, as the EP considered the Food Information to Consumers Regulation, which was to establish technical rules for package labeling of food and drink,[49] German MEP Renate Sommer, a leading opponent of nutrition profiles throughout the EP's first- and second-reading debates, introduced an amendment to remove Article 4 from the 2006 Nutrition and Health Claims Regulation. In its report on the new legislation, the Parliament's Environment, Public Health and Food Safety Committee voted to delete Article 4, sending Sommer's amendment to the full Parliament.[50] Reflecting the continued influence of producer lobbying, the June 2010 plenary vote was a 309 to 309 tie, despite substantial initial opposition to the measure. Having failed to win a majority of votes, though, the amendment was defeated by the narrowest possible margin.[51]

In addition, producers and their governments continued to contest the implications of nutrition profiling. The most publicized instance involved objections expressed by the German government to potential consequences for health claims on traditional German dark bread, which, although high in fiber content, also exceeds the salt limit proposed by the European Commission.[52] Even after the Commission wades through a range of such objections, the resulting nutrition profiles will according to the terms of the 2006 Nutrition and Health Claims Labelling Regulation be subject to the approval of the EP. In the development of the EU's regulatory regime for nutrition and health claims, only a constellation of institutional interests determined to advance European integration has counterbalanced the sustained efforts of food producers to limit the scope of the regulation.

## REGULATORY PREEMPTION BY INDUSTRY IN THE UNITED STATES: SMART CHOICES

The permissive labeling and health claims environment created by the succession of court decisions in the United States reached its logical conclusion in 2009 with the "Smart Choices" program, an industry-devised system for marking foods with a common front-of-package (FOP) icon (a green check mark) to designate their positive nutritional value. The Keystone Food and Nutrition Roundtable, a group of food producers and distributors, initiated the project in 2007, with nutritionists and consumer and public health advocates invited to participate. Smart Choices appears to have been largely an effort by food industry interests to gain the acquiescence of consumer and public health associations to the use of nutritional value as a marketing tool and to preempt steps by the FDA to consider new FOP labeling regulations.[53]

In an August 2009 letter to the Smart Choices program, the FDA expressed its commitment to regulatory oversight in spite of the succession of court judgments that have constrained its ability to regulate claims on food labels. According to the letter, research shows FOP labeling tends to steer consumers away from a review of the required nutrition facts label on the package; it is therefore "essential that both the criteria and symbols used in front-of-package and shelf-labeling systems be nutritionally sound, well-designed to help consumers make informed and healthy food choices, and not be false or misleading."[54]

A controversial element of the Smart Choices criteria concerned the treatment of added nutrients. Ultimately, the Roundtable established that foods could meet the nutrition criteria through added nutrients (in other words, regardless of the natural properties of the food substance). Michael Jacobson, executive director of the Center for Science in the Public Interest and originally a member of the panel that designed the Smart Choices criteria—and who later resigned—expressed the criticism that "[y]ou could start out with some sawdust, add calcium or Vitamin A and meet the criteria."[55] Many nutritionists were critical of the high sodium and fat content permissible within the criteria. As Lisa L. Sharma et al. report, Consumer Reports conducted an independent evaluation of Smart Choices, using its own nutrition-rating system to evaluate 27 breakfast cereals. It found many of these products to contain high levels of salt and little fiber, but especially high levels of sugar. Nonetheless, all but five of these cereals met the criteria for carrying the Smart Choices green check mark (Sharma, Teret, and Brownell, 2010: 242). Announcing plans to review potentially misleading labels, the FDA warned in its letter to Smart Choices: "If voluntary action by the food industry does not result in a common, credible approach to FOP and shelf labeling, we will consider using our regulatory tools toward that end."[56]

Given the progressive constraints on FDA regulatory authority in the sector, the warning of FDA regulatory action may have lacked the weight of a credible threat. Nonetheless, the threat of state attorneys general seeking to apply consumer protection laws against false and misleading product claims to rein in the Smart Choices program did induce producers to rethink the program. Connecticut Attorney General Richard Blumenthal announced an investigation of Smart Choices in October 2009. In response, Keystone suspended the program, and major manufacturers began to withdraw.[57]

The invocation of consumer protection laws by state attorneys general demonstrated that the same legal system that enabled the broad expansion of health claims on food labels by US food producers also constrained their latitude to craft health claims to their marketing needs. Meanwhile, in the European context, producer lobbying remained consequential even after the passage of regulation that promised to severely constrain the use of health claims on food labels as a marketing device.

**CHAPTER 4**

# RECYCLING OF END-OF-LIFE VEHICLES: "GOOD PRODUCT STEWARDSHIP" IN THE UNITED STATES; PRODUCER PAYS IN THE EU

THIS CHAPTER FOCUSES ON THE FUNDAMENTALLY DIFFERENT approaches to the management of product waste streams in the United States and the European Union (EU). The guiding concept underpinning US regulation, that of "good product stewardship," calls upon not only producers, but all actors associated with production and consumption, to share responsibility for the safe, environmentally friendly disposal of the products they handle (Toffel, 2002). In contrast, regulation developed in the EU dating back to the late 1990s is predicated on the "producer pays" concept. The fundamental idea behind this approach is that regulation must create incentives for producers to design products that incorporate fewer hazardous materials and that lend themselves more readily to environmentally sound disposal.

This difference in regulatory approaches is not a result of culturally distinct conceptions of the environment, perceptions of risk, or attitudes toward the respective roles of markets and governments. For the policy area at the center of the chapter, treatment of vehicles reaching the end of their useful lives ("end-of-life vehicles," or ELVs), national systems of regulation in European countries, though varying across numerous dimensions, were largely based on the principle of shared responsibility prior to the emergence of an EU-level regulatory regime in 2000. The EU regime grew out of a mid-1990s commitment to the producer-pays principle in the management of waste streams generally, and the institutional interaction between a European Commission committed to enacting this principle through mandatory European-level regulation, a sufficient number of national environment ministers supportive of the producer-pays approach, and a European Parliament (EP) that on balance identified its reputation as a protector of the European environment as essential to its institutional interests.

At the same time, industry opposition to a departure from shared responsibility persisted as the EU regulatory process advanced; at least one member state government—Germany, home of the EU's largest national auto-producing sector—opposed the EU directive crafted by the European Commission and supported by most other national governments. In addition, industry lobbying produced intensive debate in the EP and gained substantial support for a retreat from the commitment to producer responsibility ultimately agreed by national environment ministers interacting in the Council of Ministers. Industry advocates did succeed in winning some critical changes to elements of the directive in the final policy-formulation phase involving negotiation between the Council and the Parliament. Additionally, as in the case of nutrition labeling, industry sought, with considerable success and support from government in some national cases, to minimize their cost burden in the regulatory implementation phase.

The consequence of the "good product stewardship" model for vehicles reaching the end of their useful lives in the United States is a series of voluntary partnerships between industry, environmental groups, and regulatory authorities to dispose of portions of the hazardous waste stream, such as mercury switches in automobiles. As we have seen for the US chemicals sector, industry supports such arrangements as a means of preempting more restrictive compulsory regulation; while the results may be suboptimal, environmental groups attain some of their goals; regulators avoid costly struggles to develop and implement rules.

In the EU, institutions including the European Commission and the Parliament have backed compulsory regulation as a means of advancing European integration.

Indeed, regulation according to the producer-pays principle has yielded for the EU many of the mobilization benefits of regulation, including intensified European-level interest aggregation of producer and environmental lobbies; development of an EU-wide network of ELV recycling facilities; and adoption by all member states of legislation restricting the use of certain dangerous heavy metals in automobile production. At the same time, auto producers, particularly those who have had a substantial share of the EU market for many years and bear responsibility for the recycling of a large volume of older cars on the road, have fought to soften the impact of regulation on their ability to compete. Industry lobbying shaped the position of the German government, intensifying discord within the Council of Ministers, and is highly evident in debates of the EP. Ultimately, structure rather than culture seems to be the vital determinant of policy positions, since the call to shield producers from full responsibility for product waste streams is especially intense in the largest auto-producing countries, particularly Germany, France, the United Kingdom, Spain, and Italy.

## Producers and Responsibility for Product Waste Streams: Shared Stewardship versus Producer Pays

In the late 1980s and the early 1990s, EU institutions, including the European Commission, Council, and Parliament, embraced the concept of producer responsibility for several types of waste, including packaging waste, waste oils, batteries, and polychlorinated biphenyls (PCBs). In November 1996, the EP urged the Commission to propose legislation regulating waste streams. The EP's resolution specifically called upon the Commission to incorporate the principle of producer liability as the foundation of regulation governing ELVs. In its 1996 communication on waste management strategy, the Commission announced a departure from past waste management policy; rather than permitting producers to externalize the costs of disposal of products, these costs would now be internalized, with producers bearing responsibility for their products at the end of their useful lives (Commission, 1996).

In addition to ELVs, the EU waste management strategy addressed e-waste in the 2002 Waste Electrical and Electronic Equipment (WEEE) directive. The directive, which applies to computer equipment, household appliances, medical devices, and electrical and electronic tools, among other product categories, requires the establishment of collection systems separate from unsorted municipal waste in which consumers can return their e-waste at no charge, along with incentives for product design to facilitate dismantling, recovery, and recycling.[1]

The concept of extended producer responsibility (EPR) emerges from a critique of traditional regimes for regulating waste in which disposal and environmental impact costs of products are socialized, with local governments and taxpayers bearing this expense (Sachs, 2006: 56). Moreover, since waste disposal costs are funded out of general tax revenue rather than tied specifically to each unit of waste production, neither producers nor consumers face incentives to incorporate disposal costs into their respective production and consumption decisions (Short, 2004: 1220; Sachs, 2006: 57). The focus of EPR, then, is to alter this relationship and incentive structure, so that producers bear responsibility for their products for their full life cycle (Salzman, 1997: 1270). Or, as Sachs describes the significance of extended producer responsibility, "EPR extends the Polluter Pays Principle—which is most often discussed in the context of factory emissions, effluents, and hazardous waste clean-ups—to products themselves" (Sachs, 2006: 62).

Charging producers with responsibility for the environmentally sound disposition of their products at the end of their useful lives induces critical changes in relationships between producers and upstream and downstream actors in the product life cycle. These dynamics ensue as manufacturers consider a broader range of production decisions, including questions of product weight, recyclability, and design for reuse and disassembly (Salzman, 1997: 1273; Sachs, 2006: 64). EPR, in other words, creates strong incentives for technological innovation in material use and recovery, as well as product design (Salzman, 1997: 1274; Toffel, 2003: 107; Kibert, 2004: 511; Sachs, 2006: 52). As one set of authors describes this incentive structure, "When lifecycle environmental costs are required to be paid by the producer, implementing green production processes makes economic sense" (Nicol and Thompson, 2007: 229).

Germany's 1991 Packaging Ordinance was among the very first waste stream regulations built on the foundation of producer responsibility (Kibert, 2004: 511; Konz, 2009: 434). Driven by the problem of landfill shortage, the ordinance required producers to either individually take back their packaging or to join an industry-level take-back program (Short, 2004: 1223). The undesirability of the packaging take-back served as an incentive for producers to escape this requirement by imposing on their suppliers creation of an industry-wide recovery system. This in turn generated an incentive for suppliers to reduce the weight of packaging waste and to shift to more easily recyclable packaging materials (Salzman, 1997: 1273; Sachs, 2006: 69). The success of the German approach led to emulation by other EU member states, including France in 1992 and then Austria, Sweden, Denmark, Finland, Ireland, Netherlands, and the United Kingdom (Salzman, 1997: 1274; Short, 2004: 1226). The adoption of waste packaging regulatory regimes in several member states, along with problems created by the cross-border movement of packaging waste, created an impetus for the EU's waste packaging directive, proposed by the European Commission in 1992 (Golub, 1996: 317). While the EU waste packaging measure did not fully impose producer responsibility in practice, it did advance the principle[2]; as Sachs asserts in his discussion of the directive, "EPR as practiced in Europe has not meant that responsibility for products rests solely with producers, but European EPR does retain the core concept that producers' environmental responsibility for products extends beyond the factory door to the post-consumer stage" (Sachs, 2006: 71).

The US Environmental Protection Agency (EPA) acknowledges on its "Product Stewardship" website that "[i]n most cases, manufacturers have the greatest ability, and therefore the greatest responsibility, to reduce the environmental impacts of their products."[3] Nonetheless, shifting from this assertion to a model of shared responsibility, the EPA indicates: "[R]eal change cannot always be achieved by producers acting alone; retailers, consumers, and the existing waste management infrastructure may have to pitch in for the most workable and cost-effective solution."[4] Good product stewardship based on shared responsibility prevails largely as a means of minimizing the costs accruing to industry from efforts to reduce product waste; it is a solution to the trade-off between regulatory objectives and industrial competitiveness that predominates in the US regulatory environment. However, shared responsibility undermines the incentives generated by producer responsibility, since the environmental costs of placing the product into circulation are diffused across multiple actors (Toffel, 2002: 6). The remainder of the EPA site on product stewardship endorses cooperation between local governments and industry to encourage product recycling, and the use of incentives to good product stewardship through local government procurement policies.

Two elements of the US policy legacy help account for the lack of emphasis on producer responsibility in federal environmental law. First, from the inception of major environmental regulation in the United States following the 1970 National Environmental Policy Act and creation of the Council on Environmental Quality (CEQ), policy has focused on regulating emissions of industrial sources of pollution rather than the sorts of externalities from products addressed by EPR (Sachs, 2006: 53; 88). Second, while regulation of air and water pollution became federal responsibilities, the 1976 Resource Conservation and Recovery Act placed responsibility for solid-waste collection and disposal with the states.[5] While there are state programs covering take-back of specific products like automobile oils and beverage bottles, the waste stream life cycle generally is governed by the concept of "good product stewardship," which shares responsibility across all parties involved in the product life cycle—including manufacturers, consumers, and local governments (Short, 2004: 1227; Konz, 2009: 451).

Demonstrating that the absence of federal producer responsibility laws is not culturally determined, there have been numerous legislative efforts to move toward forms of producer responsibility in the United States. The National Recycling Act, which in the early 1990s would have imposed an obligation on producers to develop markets for recycled materials, was defeated by intensive industry opposition (Konz, 2009: 451). Around the same time, the Resource Conservation and Recovery Act (RCRA) Amendments of 1991 sought to impose recycling targets on industry, with minimum recycled product content requirements to go into effect if industry did not meet the specified targets (Keane, 1992: 262). Producers opposed these measures as prohibitively costly, presenting arguments very similar to those made by European industry in debates about EU producer responsibility initiatives (Keane, 1992: 265). At the same time, recycling industry representatives pushed for producer responsibility laws, which would have the effect of creating a national market for recycled materials.[6]

The EPA is primarily concerned with protecting its reputation for efficiency and for retaining its capacity—vital given its resource limitations—to secure voluntary agreements with industry. This explains why the EPA Administrator found provisions of the RCRA Amendments bill "too expensive to implement" and expressed a preference for reliance on market-based incentives and "voluntary initiatives on the part of EPA and industry" (Keane, 1992: 269). Guided by a "comparative risk assessment model," the EPA was in effect operating under a regulatory budget constraint.[7] Furthermore, with only a small number of states adopting specific e-waste recycling laws (California and Maine), Congress did not confront pressures to regulate at the federal level in order to create uniform conditions for industry across the national market. In 2003, a Democratic member of Congress introduced a National Computer Recycling Act to develop a national e-waste recycling program; the proposal died in committee, but was reintroduced in 2005 and again in 2007. During this period, Maine (2004) and California (2003) both enacted their own e-waste laws; the Maine law is based on the producer responsibility principle, while the California law is not.[8] When small states such as Maine act, there is little pressure from industry for the US Congress to seek regulatory harmonization; when California regulates, the result will be sufficient market incentive for some industries simply to produce to California standards—David Vogel's well-known "California effect." And in the instance of e-waste, industry lobbying was in any case successful in preventing the establishment of producer responsibility in California (Bergner, 2004).

Rather than mandatory regulation, what emerged at the national level in response to the e-waste debate was a voluntary framework prompted by manufacturers in the Electronics Industries Alliance (EIA), an outcome similar to the pattern operative in the chemicals sector discussed in chapter 2. This effort to preempt federal regulation produced the National Electronics Product Stewardship Initiative (NEPSI), an FDA-funded forum for electronics manufacturers, public authorities, recycling firms, and environmental groups to discuss mechanisms to increase collection, reuse, and recycling of used electronic equipment and encourage improved product design to facilitate e-waste reduction. The NEPSI forum initiated in 2001 culminated three years later with a resolution that recommended implementation of a national e-waste system based on an advanced recovery fee paid by consumers, with firms able to opt for their own "alternative stewardship plans," and transition to a "partial cost internalization system" (to create some incentive for manufacturers to pursue more environmentally sound product design) at an unspecified future date.[9]

Finally, discussions of EPR in the United States also took place through the Clinton Administration's Presidential Council on Sustainable Development (Sachs, 2006: 89). In this context, the Presidential Council accommodated industry's aversion to direct responsibility for waste streams by adopting a concept of "extended *product* responsibility," in which all actors in the product chain share responsibility for ameliorating the environmental impact of product waste. As Sachs asserts, "[T]he rhetoric of Extended Product Responsibility continues to dominate discussions about product externalities in the United States, disfavoring waste policy solutions that impose particular take-back or other responsibilities on producers" (Sachs, 2006: 90).

In their study of refrigerator waste comparing the EU's 2002 Waste Electrical and Electronic Equipment directive with United States and Canadian regulatory approaches, Scott Nicol and Shirley Thompson cite the failure of North America product waste stream policies to impose on producers significant responsibility for waste reduction, recycling, or environmentally sound disposal (Nicol and Thompson, 2007: 227). These authors place product stewardship on the low end of a continuum of increasing environmental protection, with shared responsibility, producer responsibility, and extended producer responsibility representing successively higher levels of producer commitment (Nicol and Thompson, 2007: 228). Regulatory approaches involving shared responsibility eliminate the benefits of internalizing environmental impact costs, "providing no feedback to the producer regarding lifecycle management costs of their products" (Nicol and Thompson, 2007: 230).

Japan's system of EPR for product waste streams is instructive in comparative context because it seeks to capture the environmental benefits of incentives for producers to improve product design while sharing costs. Motivated by similar environmental concerns to those operative in Europe, including landfill scarcity, toxic waste, and hazardous emissions from the waste management process, the regulatory regime evolved from voluntary programs to a series of materials cycle and waste management framework laws as well as sector-specific laws governing home appliances, containers and packaging waste, and construction materials, as well as ELVs (Ogushi and Kandlikar, 2007; Togawa, 2009: 273). Japan's 2003 ELVs regulation emulates the EU model—and is in part a response to EU standards that seeks to keep Japanese auto makers competitive in the European market (Togawa, 2005: 282)—in that it is designed to provide incentives for producers to consider recycling needs in the product design process. At the same time, Japan's end-of-life recycling laws are uniformly based

Table 4.1 Contrasting models of regulation of waste streams

| **Regulatory jurisdiction** | **Regulatory concept** | **Implications for producers and for the environment** |
|---|---|---|
| EU | Producer pays | Compulsory regulation in which producers bear costs of disposal of end-of-life products; strong incentives for eco-friendly product design |
| United States | Good product stewardship | Voluntary industry partnerships preempt compulsory regulation; producers do not bear costs of product waste stream, so no incentive for more environmentally sound design |
| Japan | Extended producer responsibility | Producers responsible for improving product design and recyclability; costs of environmentally sound waste amelioration shared between producers, consumers, and government |

on the notion of shared responsibility (between citizens, industry, and various levels of government) rather than the producer-pays principle. Consumers turn in end-of-life products to retailers, who in turn pass them along to producers. Manufacturers bear responsibility for environmentally sound disposal of their end-of-life products in accordance with regulatory requirements and recycling rate targets (including, in the case of ELVs, escalating requirements for recycling of automobile shredder residue, which contains toxic wastes). The costs of collection, transportation, and recovery of end-of-life products are borne by consumers, who pay a recycling fee at the time of purchase (Ogushi and Kandlikar, 2007).

Table 4.1 illustrates these contrasting models of waste stream regulation.

## European ELV Regimes Prior to EU Regulation

In the years immediately preceding the original end-of-life vehicle proposal at European level in 1997, most EU member states developed national-level ELV regimes. While in some respects these regimes were precursors to and necessary steps toward EU regulation, national efforts typically were industry-centric, involved systems of voluntary or self-regulation, and relied on markets and shared responsibility across actors rather than the producer-pays principle—in all of these respects, resembling the regulatory landscape in the United States.

Approximately 9 million vehicles reach the end of their useful lives annually in EU countries. While the metallic portion of each vehicle—about 75 percent—is fully recyclable, the remainder is shredded, generating about 2 million tons of landfill annually.[10] Automobile-shredder residue contains heavy metals (lead, cadmium, chromium) and fluids (along with plastics, glass, rubber, and foam) that constitute approximately 10 percent of the hazardous waste in the EU (European Parliament, 1999; Zaboli et al., 2000: 12).

Treatment of ELVs in Germany evolved in the context of the 1991 Toepfer Law (after Federal Environment Minister Klaus Toepfer), which introduced the producer-pays principle to the management of waste streams (Zaboli et al., 2000: II.17; Toffel, 2003: 111; Short, 2004: 1223). Amidst this emergent emphasis on producer responsibility, Germany's federation of auto manufacturers (the Verband der Automobilindustrie [VDA]) established a working group along with plastics and steel manufacturers as well as auto shredders in order to develop a response based on shared responsibility. The voluntary scheme would promote recycling by giving the last owner a financial incentive to turn the ELV into an authorized dismantler through relief of ownership taxes. The regime would rely on the market mechanism, according to which the last owner would receive or pay for recycling based on the market value of the ELV, would involve efforts to reduce the volume and toxicity of auto shredder residue (ASR), and would include a commitment by auto manufacturers to produce more readily recyclable vehicles (Zaboli et al., 2000: II.17–18; Lucas, 2001: 14–15). A viable system demanded technical regulation of dismantling and shredding processes, which the Federal Environment Ministry (Bundesministerium für Umwelt, or BMU) provided in the 1994 Altautoverordnung. In spite of the establishment of technical regulations, the Environment Ministry and auto producers maintained different perspectives on the most effective approach to ELVs, with the ministry favoring producer responsibility and industry preferring self-regulation (Zaboli et al., 2000: II.19).

While the initial proposals of the working group excluded any quantitative targets, the VDA's position on the issue evolved over time; in addition, manufacturers created incentives for materials producers by signing contracts with those who agreed to organize nationwide environmentally sound ELV processing networks. To preempt formal regulation on the producer-pays model, the VDA along with other professional associations—car dismantlers; the steel recycling industry; scrap recyclers; metal traders; the rubber industry; plastics producers and processors; auto-parts traders—submitted to the BMU a voluntary plan for ecologically sound disposal of ELVs, which included a nationwide collection infrastructure; development of environmentally sound processing procedures for fluids, dismantling, parts recovery, and recycling; and numerical recovery targets to reduce ASR. In addition, acknowledging the BMU's commitment to the producer-pays principle, the plan allowed for free take-back of ELVs younger than 12 years from first registration (Zaboli et al., 2000: II.19–20; Lucas, 2001: 15).

The Federal Economics Ministry endorsed an ELV scheme on these terms, including a voluntary pledge by industry to take back ELVs according to the market mechanism (i.e., based on the scrap value of the ELV). The regime was approved by the German legislature in 1997, and the German government put in place supportive regulatory measures governing the authorization of collection centers and the obligations of shredding and recycling facilities (Zaboli et al., 2000: II.20–21). The coming into force in 1998 of a voluntary regime endorsed by the government and so consistent with the vision of the auto industry adds to our understanding of the especially staunch opposition of German auto producers (detailed below) to the more demanding EU regulatory regime proposed in 1999 (Zaboli et al., 2000: II.22).

The UK case provides further evidence of industry support for shared responsibility rather than producer responsibility, as well as the willingness of government to endorse this principle. Britain developed a voluntary system involving auto producers, suppliers of materials and parts, and ELV shredders and dismantlers, as well as the Department of Industry and the Department of Environment (Zaboli et al., 2000: II.64). The program, "Automotive Consortium on Recycling and Disposal" (ACORD) was finalized in 1997 and focused on developing an ELV treatment system, with particular emphasis on reducing landfilling of automobile-shredder residue.

ACORD was designed as a market-based voluntary system, with government playing only a consultative and monitoring role; one objective was to prevent "overregulation" from the EU level (Zaboli et al., 2000: II.64–65). ACORD set target recovery rates for automobile-shredder residue, with targets to be attained through a combination of recycling and energy recovery (Zaboli et al., 2000: II.65). Auto manufacturers accepted responsibility for producing cars more suitable for recycling; dismantlers would remove a larger share of nonmetallic materials and separate materials more effectively before delivery to collectors and shredders. Rather than any mandatory system, last owners were to be encouraged to deliver ELVs to certified dismantler sites.

In 1993, France established a national voluntary agreement between the government and large producers, dismantlers and recyclers, and material manufacturers, known as the Accord Cadre (Zaboli et al., 2000: II.2). The agreement included targets for waste reduction and ASR levels; auto producers pledged to increase the use of recycled materials, though they did not commit to significant investments or technological innovations. Recycling firms agreed to pursue technical and efficiency standards, while public authorities promised to combat unauthorized dumping of ELVs (Zaboli et al., 2000: II.3).

Seeking to reduce a growing number of abandoned vehicles, the Swedish government introduced an ELVs regime as early as 1975. Evolution of the Swedish regulatory environment again illustrates the insistence of producers on shared responsibility for the waste stream, although in this case the government moved toward a commitment to the producer-pays principle in the early 1990s.

Sweden's original system required that car buyers pay a vehicle scrapping fee at initial purchase; the proceeds went to a public fund to cover payments to the final owner upon delivery of the ELV to an authorized treatment facility and also to the dismantling firm (Zaboli et al., 2000: II.55). In 1993, the government sought to introduce technical regulations for recycling, higher quantitative targets, and free take-back from the ELV owner based on the concept of producer responsibility. Swedish auto producers, viewing the ELV matter similarly to their counterparts in other European countries (and, for that matter, in the United States), were strongly opposed to this approach and proposed instead a voluntary scheme based on cooperation between actors in the waste stream and on shared responsibility. Nonetheless, the Swedish government persisted in the context of an overall commitment to producer responsibility that extended to packaging and waste paper and vehicle tires,[11] and in 1997 enacted its "Ordinance on Producer Responsibility for Vehicles" based on the producer-pays concept (Zaboli et al., 2000: II.57).

## ELVs in the United States: The Role of Markets and Industry Preemption

Approximately 11 to 12 million vehicles reach the end of their useful lives in the United States annually (Staudinger and Keoleian, 2001: 3; 5). Five million tons of ASR are landfilled each year.[12] As with waste streams regulation in the United States generally, these activities are not regulated at the federal level. Instead, they are governed by two elements: first, market relationships; and second, as indicated by Raymond Konz, "individual States are free to adopt inconsistent regulations, or forego regulation altogether" (Konz, 2009: 432).

There has been only a single piece of legislation introduced at national level focused on treatment of ELVs: the 1991 Automobile Recycling Study Act (Staudinger and Keoleian, 2001: 36). The act, which did not progress beyond the Subcommittee on Transportation and Hazardous Materials of the House Committee on Energy and Commerce, would have required the EPA to undertake a study of how best to increase recycling of auto components; promote substitution for toxic materials; and encourage attention to environmentally sound disposal in product design (Staudinger and Keoleian, 2001: 36). In the absence of such legislation, regulation of ELV disposal has been limited to a patchwork of state laws governing vehicle fluids, batteries, and landfill restrictions (Short, 2004: 1227).

Voluntary and market-based schemes are prominent. Auto producers have engaged in limited efforts to promote recycling of hazardous metals and have entered into partnerships with government departments and other industrial sectors to enhance vehicles recycling without imposing costs on manufacturers, driven at least in part by a desire to preempt federal regulation.[13] The National Vehicle Mercury Switch Removal Program was initiated in 2006 by the End of Life Vehicle Solutions Corporation, an entity founded by major auto producers in 2005 in accordance with state governments, scrap metal recyclers, and the EPA.[14] The program to collect and recycle mercury switches in ELVs to prevent these from entering shredder residue is the first and only program instituted by End of Life Vehicle Solutions. Since the use of mercury in switches in new vehicles was phased out in 2002, the mercury-switch program had no impact on auto design.

Arrangements governing electronic equipment waste streams in the United States replicate those prevailing for ELVs. In July 2011, the CEQ, the EPA, and the General Services Administration (GSA) released their "National Strategy for Electronics Stewardship," the output of an interagency task force created by executive order in November 2010 to address the problem of unsafe handling and disposal of a mounting volume of waste—estimated at 2.5 million tons annually—from electronic technologies.[15] The strategy seeks to create incentives for design of greener electronics and to encourage manufacturers to expand take-back programs and to work with certified recyclers.

The mechanisms at the core of the program for electronic equipment stewardship are entirely market-based and voluntary. These include stakeholder dialogues; funding for intensification of research and development to improve products; certification and information programs to promote consumer purchases of green electronics; and, perhaps of most consequence, use of federal government market power to lease and purchase greener electronics, including large-scale Department of Defense (DoD) electronic weapons systems and the high-volume electronics purchasing of the GSA. While several of the largest electronics manufacturers—Dell, Sprint, and Sony—announced their agreement to voluntary partnerships, the program nonetheless imposed no mandatory burden on producers.

## Regulation of End-of-Life Vehicles in the EU

The European Commission's 1997 ELV proposal was drafted by the Environment Directorate General (DG) and incorporated the producer responsibility principle for new as well as existing vehicles (Commission, 1997). In addition to requiring elimination of lead, mercury, hexavalent chromium, and cadmium in vehicle production by July 1, 2003, the proposal set a reuse and recycling rate of 85 percent for new vehicles beginning in 2005 and 95 percent beginning in 2015 (Commission, 1997: Article 7).

As discussed above, numerous company-level or industry-sponsored schemes and industry-government partnerships emerged across EU member states in the years just prior to the Commission's draft ELV directive. European producers tended to act preemptively—in this sense coinciding with the behavior of US industry; as Zaboli et al. point out: "Expectations about regulation strongly shape the whole innovation process in ELV. Innovation initiatives and achievements occurred before the most important regulations at EU and national level were introduced" (Zaboli et al., 2000: 16, I.38). At the end of 1999, prior to final agreement on the ELV directive among EU institutions, 10 of 15 EU countries had in place ELV regulations or voluntary agreements, with industrial agreements in process in three additional countries.[16] These industry-led regimes tended to be based on the principle of shared responsibility, as opposed to producer liability. The arrangements typically excluded free take-back of vehicles at the end of their useful lives; placed no limits on combustion of ASR (though ASR contains high levels of heavy metals, chlorine, and PCBs); and also excluded limits on the use of heavy metals (Zaboli et al., 2000: 13).

While manufacturers did not oppose the introduction of producer responsibility and an increase in reuse and recovery shares for newly designed autos, they attacked the retroactive nature of the proposed EU legislation. ACEA, the European Automobile Manufacturers' Association, estimated the cost of taking back each scrap auto at 370 Deutschmark, nearly $200.[17] With 160 million vehicles already on the road, European auto manufacturers warned that the costs of the proposed regulation would harm their ability to compete with non-European producers—both recent entrants to the EU market with few vehicles on the road and those not facing a similar regulatory burden in their home markets.

Auto manufacturers pressed the Commission to address producer concerns in the draft proposal. Having met with little success at this stage in the legislative process, manufacturers sought to line up opposition to the regulation in the Council of Ministers (Coen, 2004). German car manufacturers, who produce two-fifths of Europe's autos, sought to enlist the Federal Economics Ministry in efforts to alter the proposed regulation—by distributing the cost burden between producers and consumers—over the heads of national environment ministers.[18] Organized environmental interests, such as the European Environmental Bureau, lobbied heavily for the retention in the regulation of the "producer pays" principle.

There are several plausible hypotheses concerning the determinants of the behavior of the German government in the Council of Ministers during this period. Rüdiger Wurzel considers two possibilities: either a Germany more willing than in the past to assert national interests, or a Germany governed by a red-green coalition more likely to advance strong environmental positions (Wurzel, 2000: 24–25). Studying the determinants of government positions in Council of Ministers negotiations, Stephanie Bailer finds evidence that contradicts literature pointing to left-right partisanship as the driving element. Instead, she finds that the positions of governments in day-to-day Council negotiations are determined by the structure of domestic interests rather than partisanship or strategic calculation. The use of strategic bargaining is undermined by the "information-rich environment" of Council of Ministers negotiations, in which bluffs are likely to be detected and negotiating leverage thereby undermined (Bailer, 2008: 8). For Germany's Social Democratic–led government, determination of bargaining position by the structure of domestic interests meant support for the powerful auto industry, even if this implied an "anti-environmental" position (Bailer, 2008: 6).

While national government environment ministers reached agreement on the outlines of a common position in December 1998, the Council, under the German presidency during the first half of 1999, nonetheless decided in March 1999 to postpone final agreement until June in response to a German government request for additional time for consultation with national auto producers.[19] Auto-industry lobbying was highly effective, and the June meeting of national environment ministers was unable to reach resolution as promised by the German environment minister when he requested a postponement in March. The draft agreement now faced a blocking minority—with Germany supported by the United Kingdom and Spain (Wurzel, 2000: 34)—though the Council statement indicated that an agreement might be reached by delaying the entry into force of the requirement for manufacturers to take back vehicles already on the market.[20]

The Council ultimately adopted its common position in July 1999, following a transfer of the Council presidency to the Finnish government, which made ELVs directive a priority. The backtracking of the German environment minister from the December 1998 agreement "poisoned the atmosphere" of the first Environmental Council meeting under the German presidency and was viewed by Council members as a violation of Council presidency's "honest broker" norm (Wurzel, 2000: 30, 34). National environment ministers also were angry about the logrolling between the German and UK governments that won over UK support for the German position.[21] Accordingly, Finland's environment minister was able to broker a deal even with Germany voting against the common position and the Netherlands abstaining.[22] The deal focused on the date from which producer responsibility for ELVs would begin; the common position set the dates at January 1, 2001, for vehicles placed on the market from that point forward and January 1, 2006, for vehicles on the market prior to January 1, 2001. Despite German opposition, the Council's July 1999 common position adhered to producer pays for new and existing cars.[23]

Ironically, the European Parliament, with its long-standing commitment to environmental concerns, moved closer than the Council to endorsing a system of shared responsibility for ELVs. The argument that structure more than partisanship determines bargaining positions in the Council of Ministers appears to apply to members of the EP as well, as illustrated by the coincidence of positions of the leading German member of the European People's Party (EPP) and the leading member of the Party of European Socialists (PSE). Karl-Heinz Florenz, a German Christian Democrat (hence a member of the EPP) and rapporteur for the Committee on Environment, Public Health and Consumer Protection, and Bernd Lange, German Social Democrat (PSE) and member of the EP's Environment Committee (ENVI), agreed that producers should be spared the costs of taking back all vehicles already on the road at the end of their useful lives. The principal difference in their positions was that the Christian Democrat argued for cost sharing between producers and consumers, while the Social Democrat called for a public fund to spare consumers recycling costs. In addition, Lange proposed an amendment to the Council's common position that would involve a delay in the application of producer responsibility for vehicles already on the market.

Speaking on behalf of the EPP during the Parliament's second reading of the ELV bill, Florenz argued that manufacturers will simply pass along costs to consumers. In addition, he suggested these costs will be higher than necessary because manufacturers will have a monopoly on recycling. Finally, he indicated that the retroactive nature of the take-back requirement would likely invite a legal challenge. A fund generated from recycling costs paid by new car buyers, along with matching contributions from manufacturers, would enable take-back of ELVs from the last owner at no cost beginning in 2006.[24]

PSE spokesman Bernd Lange invoked the example of Britain's Rover—a repeatedly dissolved and restructured company with nonetheless identifiable links to a long prior production history—to make a case, based on the impact on competitiveness, against retroactive manufacturer responsibility for all cars on the road. Noting that Rover would bear responsibility for 5.8 million vehicles while a recent market entrant from South Korea would have very little liability, Lange argued: "Such distortions in competition are not about environmental protection; they simply affect investment potential and the jobs of those building cars in Europe."[25] Lange additionally called upon colleagues in non-auto-producing countries "to show solidarity with the more than 2 million people who make a living from building cars in Europe, so that these jobs can be secured in the future."[26]

One British PPE member echoed the positions of Florenz and Lange on the need for cost sharing and opposition to retroactive responsibility of manufacturers for their vehicles on the road: "That would mean billions of pounds or euros for each of the major companies in each of the countries of the European Union... Because European car manufactures have been operating here for many, many more decades than companies from Japan, Korea and elsewhere, that would be a much greater burden on the older European companies and a competitive gift to their competitors from elsewhere."[27] Another British Conservative member of the PPE similarly rejected the EU directive outright on economic grounds: "[I]t dumps huge costs on the European motor industry which would damage competitiveness and damage employment. In this Parliament we constantly talk about the need to promote employment and jobs in Europe and yet we constantly pass measures which will have the effect of reducing employment."[28]

Adding to the evidence in support of structural rather than partisan determination of positions of EP members, members of the PSE from France—the second-largest auto producer in the EU after Germany—endorsed Lange's proposed amendment to delay producer responsibility, while a Danish member of the PSE spoke in the EP debate in direct opposition to the amendment and in favor of producer responsibility and the Council's common position. One French PSE member argued that both the Florenz and Lange amendments "have the advantage... of reconciling environmental constraints and economic imperatives." By imposing most or all costs of ELV recovery and recycling on manufacturers, the Council's common position "is totally unfair to European car manufacturers."[29] Furthermore, endorsing Florenz's critique of retroactive responsibility of manufacturers for vehicles on the road, as well as Lange's argument regarding the potential for environmental protection efforts to distort competition, the PSE member argued, "This solution is unacceptable because these car manufacturers have not had the opportunity to build the environmental demands we are making of them today into their manufacturing procedures and their manufacturing costs. The common position puts European car manufacturers in a disadvantageous position in relation to manufacturers which have just come to the European market."[30]

In addition, Guido Sacconi, a well-known environmentalist and PSE member from Italy, another substantial auto-producing country, also endorsed the Lange approach of deferring the imposition on manufacturers of recycling costs for vehicles already on the road. The Danish PSE member, in contrast with Sacconi and French PSE members, expressed frustration with advocacy for producers from members of the Parliament's Environment Committee and urged a sustained commitment to producer liability:

> If we do reduce it, we remove from manufacturers the incentive to design and produce cars which give rise to less waste. The common position, which was arrived at with great difficulty in the Council, which the Commission has endorsed and which was also maintained in Parliament's Committee on the Environment, Public Health and Consumer Policy, complies fully with the environmental requirements. I therefore find it incomprehensible and very curious to see amendments from members of the Committee on the Environment, Public Health and Consumer Policy, the objectives of which are to weaken the environmental requirements and significantly reduce car manufacturers' liability.[31]

Unsurprisingly, PSE members largely rejected an amendment for cost sharing proposed by PPE rapporteur Florenz.[32] However, the vote tally of PSE members on the second-reading amendment of Bernd Lange, which would alter the text of the legislation to delay producer responsibility for registered vehicles until 18 months after the entry into force of the directive, provides additional evidence for a structural interpretation of voting behavior.[33] The plenary vote was 322 in favor and 199 against, with 11 abstentions. PSE members voted 110 for and 51 against. Of this group, 30 German PSE members voted for the measure, while none opposed. French members of the PSE voted 18 in favor of deferring producer responsibility, with only 2 against.

And the Italian PSE delegation voted 13 to 1 in favor of the amendment. In other words, in all of these major auto-producing countries, members of the European Socialists endorsed a delay in producer responsibility in far greater proportions than the EP membership as a whole.[34]

PSE members from the United Kingdom, on the other hand, present an anomaly, with 3 members supporting the amendment and 23 opposing. This result invites alternative interpretations of the EP's vote on amendments to the producer responsibility requirement. We may assume from a partisan perspective that members of the center-left are more likely than members of Parliament on average to endorse measures to protect the environment. In contrast with center-right members of the European Parliament (MEPs), center-left MEPs may be less responsive to industry concerns about regulatory costs. At the same time, PSE members may be interested in the economic and employment impact of regulatory burdens on industry. We would expect MEPs belonging to the EP's major committees concerned with industry and economic competitiveness—the Economic and Monetary Affairs Committee or the Committee on Industry, External Trade and Research[35]—to be especially sensitive to these priorities.

As table 4.2 illustrates, of the 110 PSE members supporting the Lange amendment, 10 were members of the Parliament's Environment Committee. Of those voting against, a very slightly larger share were ENVI members. But the difference is not dramatic. Members of the Economic and Monetary Affairs and Industry, Trade and Research committees represented a larger share of those opposing the amendment than those supporting. At the same time, while 68.3 percent of all PSE members voted to defer producer responsibility (excluding abstentions), only 58.6 percent of PSE members of the two major industry and competitiveness committees (17 of 29) supported the measure. This is not the result we would expect under the assumption that members of these committees are more responsive to industry concerns than the average PSE member.

Table 4.2 Votes of Party of European Socialists, second-reading amendment to defer producer responsibility (Members of Environment Committee versus Members of Economics and Industry Committees)

| **PSE members voting** | **For amendment to defer producer responsibility** | **Against amendment** |
|---|---|---|
| Of these | 110 | 51 |
| Members of the Environment Committee | 10 (9.1%) | 6 (11.8%) |
| Members of the Committee on Economic and Monetary Affairs + Committee on Industry, External Trade and Research | 17 (15.5%) | 12 (23.5%) |

*Source:* Figures calculated by author from voting results reported in *Official Journal of the European Communities*, 2000/C309, October 27, 2000. Available at http://eur-lex.europa.eu/JOHtml.do?uri=OJ:C:2000:309:SOM:EN:HTML

In contrast with the uniform support for the EP amendment to defer producer responsibility among German MEPs belonging to the PSE, all 5 Swedish members voted against, as did all 3 Danish members. For Greece, widely considered an environmental protection laggard—but a nonproducer of automobiles—all 9 PSE members opposed the amendment. For Spain, which has a sizeable auto sector, 22 PSE members endorsed the amendment; only 1 opposed. The 22 included both Spanish PSE members on the EP's Environment Committee.

A possible explanation for the lack of support for delaying producer responsibility among the PSE delegation from the UK concerns efforts by the new Labour government to establish its environmental credentials within the EU. Wurzel, for example, notes that while in the Council presidency in 1998, the British government sought to restore its reputation as a European partner and to distance itself from its image as "dirty man of Europe," and that German Chancellor Schröder's request for support for the German position was not especially welcomed by the Blair government (Wurzel, 2000: 34).[36]

However, while these data support a structural account of the voting behavior of MEPs, political partisanship was evident in the voting pattern of Liberals and Greens. Members of these parties—including MEPs from major auto-producing countries—uniformly rejected the Florenz amendment to share costs; the Lange amendment to defer producer responsibility met with nearly unanimous opposition from Liberals and Greens.[37] Both Liberals and Greens were central to the EP's role in sustaining the producer responsibility principle in the face of heavy industry lobbying.

Members of both parties underscored the point that producer responsibility is essential as an inducement to more environmentally friendly design of automobiles. Most critically, they made the case that the EP's reputation and role as a protector of the environment was at stake.

During the second-reading debate, a Dutch member of the European Liberal Democrats (ELDR) applauded the balance struck by the Council in its common position and argued that since most ELVs have some market value, the increase in recycling of parts associated with the producer responsibility requirement would be a boon to the small and medium companies in the recycling sector.[38] Arguing against both the Florenz and Lange amendments, a Dutch Green suggested that the results of the EU's waste packaging directive demonstrate that shared responsibility does not work. Additionally, assigning producers full responsibility from 2006, as established in the Council's common position, gives auto producers sufficient time to prepare for full responsibility; the Lange amendment to delay producer responsibility should therefore be rejected. The Green MEP concluded with a case for the role of the EP as an advocate for the environment: "For twenty years, the majority within Parliament has attempted to make European environmental policy greener. Today, this green position is being threatened under pressure from, in particular, the German and French car industries. That is why I urge you to vote against Amendments Nos. 38 and 45."[39]

A German Green echoed the call to defend the Parliament's environmental role, while implicitly attacking the position of the German government (the fifteenth member state): "Will we give our backing to forward-looking environmental and consumer protection or, as the amendments of Florenz, Lange and others cause me to fear, will we allow ourselves to become a servant of the German car industry? The credibility of Parliament as a champion of environmental protection is at stake. It would be shameful to say the least if the European Parliament were to draw back from the professions the governments of the 14 Member States and the European Commission have been making concerning manufacturer responsibility and environmental protection!"[40]

A French member of the Greens also argued that the EP's credibility as a protector of the environment and the concerns of citizens was at stake: "Today, if 314 Members of Parliament yield to the strength of the lobbying of some car manufacturers, we should be doing ourselves dishonour. In expecting consumers to bear half the cost of the recovery of end-of-life vehicles, as Amendment No. 38 stipulates, even though the common position states that all recovery costs should be borne by the manufacturer, Parliament would, for the first time, not be acting as the defender of the rights of the consumers and citizens of Europe, it would be turning into mere sounding box for different lobbies."[41]

A Swedish member of Nordic Green Left also cast the proposed amendments to the Council's common position as a test of the environmental commitment of the EP:

> With this directive, we have the chance to take a big step forward, but this presupposes that the Council's position is not torn to shreds and weakened in Parliament's reading of it. If they were adopted, many of the amendments which have been tabled would considerably weaken the directive. This applies, above all, to amendments from the Group of the European People's Party but also, unfortunately, to some of the amendments from Mr. Lange... If those amendments were adopted which would considerably weaken the directive, this would be very detrimental, not only from an environmental point of view but also for the European Parliament's credibility on environmental issues.

> Reference was made earlier on in the debate to the fact that some thought should be given to the millions of people who work in the car industry in various countries and to those countries which have large car industries, for example my own country, Sweden . . . I think that very tough demands should be made of the car industry. These would, of course, be to the advantage of modern, progressive car manufacturers who think in environmentally friendly terms. It is precisely this type of car industry which we should be encouraging in the European Union.[42]

Finally, a Danish member of the PSE both reinforced the threat to the EP's integrity as guardian of the environment and warned of the implications for forthcoming waste streams legislation: "If these amendments are adopted, I think that we shall cause Parliament seriously to lose credibility on environmental issues. So far, we here in Parliament have been positive catalysts for environmental protection but now, if the amendments proposed by Mr. Florenz and Mr. Lange are adopted, we are to become a negative factor for the environment in Europe. What is more, if we reduce manufacturers' liability on this issue, this will also have serious consequences for subsequent matters in other areas, for example the forthcoming directive on scrap from computers and other electronic equipment."[43]

At the close of the EP's second-reading debate, the EU's Commissioner for the Environment, Margot Wallström, offered the European Commission's response. In doing so, she amplified the environmental role of the EP: "Parliament has traditionally made a major contribution to strengthening environmental legislation in Europe. It would amaze and depress me if it were not to do so today."[44] Wallström indicated that the Commission viewed the Council's common position as "fair but fragile," and could not support amendments to the provisions for manufacturers' liability "which bring the absolutely basic pillars of the present proposal into question." While the Commission is largely a facilitator once the Council has arrived at a common position, according to the EU's legislative co-decision procedure, the Commission's position on second-reading amendments of the EP determine the decision rule in the Council for adopting or rejecting EP amendments.[45] If the Commission accepts EP amendments, the Council can accept these by qualified majority; in the absence of Commission support, Council adoption of amendments requires unanimous support. In this instance, the Commission accepted part or all of 13 of the EP's 32 second-reading amendments. Among the amendments rejected by the Commission were measures to delay the requirement that manufacturers phase out the use of heavy metals and the Lange provision limiting producer responsibility to vehicles registered 18 months after the entry into force of the directive, as well as an amendment pushing back the timetable for requiring higher reuse and recycling thresholds.[46]

Of the EP's 32 second-reading amendments, the Council accepted only 6, invoking the EU's legislative conciliation procedure. Following informal "trialogues" with the delegations of the Council and the Parliament, the Commission submitted compromise texts that resolved most issues, with producer liability and heavy metals the persistent sticking points (European Parliament, 2000: 7). Ultimately, conciliation confirmed the principle of producer liability, but deferred liability for vehicles put on the market prior to July 1, 2002, to January 1, 2007—a delay of one year from the date of 2006 established in the Council's common position (European Parliament, 2000: 7). In addition, conciliation sustained the ban on lead, mercury, cadmium, and hexavalent chromium for vehicles placed on the market after July 1, 2003, but provides for some exemptions on technical grounds.

## Implementation of the EU Directive and the Persistence of Industry Lobbying

The ELV directive, due to take effect in April 2002, did not specify how disposal of autos already on the market before January 1,

2002, would be funded prior to January 1, 2007, when take-back of the ELV at no cost to the final owner was required. In addition, Article 12 of the directive permitted governments to move toward full producer responsibility for these vehicles before January 1, 2007. As a result, domestic interests—auto producers, recycling companies, environmental associations, and local governments—lobbied intensively over the terms of this transition period.[47]

National governments arrived at different approaches to the transition, a product of both lobbying and existing ELV regulatory systems. Reflecting its prior institutionalization of the producer-pays principle, Sweden adopted a system involving no cost to the final owner of all vehicles on the market after January 1, 1998, effective July 1, 2001—far ahead of the mandatory date of January 2007 (European Parliament, 2007: 50). The Italian government, in contrast, while putting in place legislation establishing responsibility of auto manufacturers for the creation of a network of treatment plants for the free take-back of vehicles by 2003, did not require the Ministry of the Environment to issue implementing measures until 2006. This meant there would be no take-back of any vehicles at no cost to the final owner (including those entering the market after January 1, 2002) until January 1, 2007—one of several bases on which the European Commission referred the Italian government to the European Court of Justice (ECJ) for inadequate transposition of the directive into national law (European Parliament, 2007: 38).

In the United Kingdom, the auto industry lobby won government support for its preferred course of transposing the directive. For treatment of vehicles put on the market prior to January 1, 2002, and reaching the end of their useful lives during the 2002–7 period, options for the incidence of costs of recycling vehicles already registered included placing these entirely on manufacturers; absorption of the cost by government; taxing new vehicles to create a recycling fund; and imposing the cost on the last owner. Local councils opposed this last option, since it would create incentives for owners to abandon ELVs, adding to an expensive problem already facing local governments.[48] Environmental groups such as Friends of the Earth as well as the British Metals Recycling Association warned that imposing the recycling cost on the last owner of the vehicle would lead to large increases in illegal dumping.[49]

The chief executive of the Society of Motor Manufacturers and Traders (SMMT) argued that the government was weighing down the auto industry with "a costly and totally impractical liability." The SMMT suggested, for example, that the burden on MG Rover stemming from the cost of recycling its legacy vehicles could force the company into bankruptcy. Recycling costs should therefore be shared between manufacturers, owners, insurance companies, and the British Treasury.[50] The SMMT insisted that imposition of costs on producers should not take place until 2007—a position that prevailed in the United Kingdom's 2005 End-of-Life Vehicles Regulation, as did the industry position on implementation of the EU's WEEE Directive, which the UK government delayed from August 2004 to 2005 in order to mitigate the cost burden on business.

As the British government considered the process for implementation, the House of Commons in May 2002 debated a report on the ELV directive from the Parliament's Select Committee on Trade and Industry. Both opposition Conservative MPs and the Labour government's Industry Minister expressed a commitment to minimizing the cost burden on industry. The most critical question concerned "accrual"—historic versus current market share—and whether costs could be distributed based on market share, with those firms that had recently expanded their share (including market entrants from abroad, such as Japanese and South Korean firms) subsidizing the historic market share of others. The chairman of the Parliament's Select Committee on Trade and Industry noted the especially difficult situation of companies like Rover, GM, and Ford with the largest "historic car parc" (accumulation of vehicles on the road) but which had diminishing market shares, implying fewer resources with which to meet the costs of producer responsibility.[51]

One Conservative MEP warned of the effect on her local MG Rover plant in Birmingham, which was struggling to reemerge and in the process of finalizing a partnership deal with a Chinese company: "We have significant liabilities in our car parc, which the present company could clearly not sustain. It would be not a severe problem so much as a knockout blow, if the company had to pick up the liabilities as the directive seems to suggest."[52] Another Conservative MEP added, "Given the overcapacity in the world automobile industry, we must try to minimise the disadvantage to manufacturers in the United Kingdom. That is not protectionism; it is common sense."[53] And still another asked: "Are the Government willing to sidestep the intention of the directive, which appears to be that producers should pay? Does the Minister agree that a methodology that avoids placing the burden on producers should be considered?" The MP added that "the alternatives could threaten the UK motor industry's competitiveness . . . and may place an intolerable strain on the balance sheets of individual companies."[54]

In response, the Labour government's Minister for Industry and Energy, Brian Wilson, assured MPs that the government would implement the EU directive "with a light regulatory touch at first . . . with the minimum disruption to current market systems and . . . without putting UK business at a competitive disadvantage to its EU counterparts." Pointing out that producer responsibility "is not an optional element" in the directive, the minister nonetheless indicated, "We want the maximum environmental benefits, but we want to minimise the cost."[55]

As chapter 3 demonstrated for the regulation of health claims on food labels and efforts to establish provisions for nutrition profiles and health function claims, even in instances in which industry lobbies have not achieved their preferred outcomes, attention to regulatory cost burdens and their impact on industrial competitiveness is deeply institutionalized at both European and national levels. We should not read differences in regulatory outcomes between the EU and the United States as evidence of fundamentally different cultural frames or the relative weakness of the industry lobby in the EU. As I discuss in the concluding chapter, while EU regulation of ELVs and health claims on food labels derive from the thrust toward integration created by the institutional structure of the EU, implementation deficits represent an equilibrium outcome to the EU's regulatory trilemma involving efforts to safeguard environment and public health, advance integration, and protect economic competitiveness.

# CHAPTER 5

# Regulatory Trade-offs and Outcomes

To assert that environmental and health regulation have become extremely rigorous in the European Union (EU) during the past decade while they have languished in the United States would be to oversimplify contrasting but variegated regulatory trajectories. Regulation in the areas of chemicals, health claims, and recycling of end-of-life autos in Europe has become harmonized at the EU level and reflects a serious commitment to reducing dangers to the environment and public health, but intensification of regulatory rigor has been attenuated by effective introduction into the policy process of a desire to limit the cost burden on industry. In the United States, efforts by some policy makers to initiate more restrictive regulation of chemicals have been preempted by voluntary programs sponsored by industry; a long history of rigorous regulation of health claims on food labels has been transcended by court decisions expanding the free exercise of commercial speech; and systematic recycling of end-of-life vehicles (ELVs) is confined to a few state laws governing portions of the recovery and recycling process, coupled with narrowly targeted voluntary industry agreements.

The principle of precaution is indeed embedded in the EU's approach to regulation, having been introduced as the guidepost for environmental policy in the 1992 Maastricht Treaty.[1] But precaution is hardly foreign to the US regulatory landscape, as we have seen from the original aspirations of the Council on Environmental Quality's mandate to regulate dangerous chemicals as well as the long history of stringent regulation of health claims on food labels. At the level of policy conception and design, there is no simple cultural divide between the United States and EU settings defined by societal willingness to tolerate risk.

Furthermore, public authorities in European countries and the EU seem no more anxious than American policy makers to advance environmental objectives at the expense of business interests and industrial competitiveness. At the national level, regulatory regimes governing the recycling of end-of-life autos were mostly voluntary and did not impose costs directly on producers prior to the emergence of an ELVs directive at the European level. Throughout the formulation and debate over that directive, industry representatives found points of access that tilted the emphasis away from the imposition of recycling costs exclusively on producers, although ultimately they were only able to delay the onset of producer responsibility. Industry associations had a more profound impact on the outcome of EU chemicals regulation, which was revised during the course of the legislative process to soften demands on industry for reporting information and substitution for dangerous chemicals. In the debate over regulation of health claims on food labels, industry interests persuaded the European Parliament (EP) to remove the most contested clause on nutrition profiles, though this provision was restored by national health ministers in the Council of Ministers.

More broadly, cost-benefit analysis and ex ante regulatory impact assessment are central to policy making in the EU, just as they are in the United States—though impact assessment does have a longer tradition in the United States, dating back at least three decades (Renda, 2006: 7; Close and Mancini,

2007: 6), while the use of regulatory impact assessment became systematized in the EU only with the launch of the "better regulation" agenda in 2001 (Renda, 2006: 48). In the United States, executive orders requiring regulatory impact analyses for major regulatory measures have transcended political partisanship, having been promulgated by both the Reagan and Clinton administrations.[2] Separate legislation calls for impact assessment of measures affecting the small business sector, also a focal point in the EU.[3]

Impact assessments take place at different points in the policy-making process in the United States and the EU. In the United States, impact assessment is used to achieve efficiency in agency rulemaking (Close and Mancini, 2007: 8). In the EU, the European Commission uses regulatory impact assessment as a tool in the process of drafting legislative proposals (Close and Mancini, 2007: 8). In addition, the Council of Ministers and EP in 2005 entered into an interinstitutional agreement with the Commission to undertake impact assessments of their substantive amendments to proposed legislation.[4] In both the United States and EU cases, the detail of impact assessment is proportional to the magnitude of the proposed regulation (Close and Mancini, 2007: 24).[5]

One of the central stories of this analysis of environmental and health regulation in the United States and the EU is precisely that both institutional innovation in all the principal policy-making bodies—including the formation in 2002 of the Competitiveness Council as a configuration of the EU's Council of Ministers; the creation of the Internal Market Committee (IMCO) in the European Parliament in 2004; and the development of an Impact Assessment Board within the European Commission—and the organizational response of industry associations to correspondingly augmented opportunities for interest articulation at the EU level, have enhanced the impact of the industry lobby on EU legislation. Overall, the intensified focus on industrial competitiveness that has characterized policy making in the EU since the launching of the Lisbon agenda in 2000 fundamentally challenges the contention that EU policy making reflects a much stronger philosophical predilection for public over private goods than policy formulation in the United States.

Looking across regulatory landscapes in the two systems, we find a high degree of protection of free commercial speech in both. European Courts have struck down laws across several countries, reasoning that the measures in question disproportionately restrict markets relative to the regulatory objectives sought. Courts in the United States similarly have overruled Food and Drug Administration (FDA) regulatory action on grounds that consumer protection objectives could be secured through less restrictive means than outright bans on health claims on food labels. Given this inclination of courts to safeguard markets from regulatory overreach, the mutual presence of precaution, and the shared concern for the regulatory cost burden imposed on industry, we would expect similar regulatory outcomes across the United States and the EU.

Nonetheless, in the United States there is a highly ineffectual federal regime for regulating dangerous chemicals, along with an industry-sponsored voluntary program pledged to reduce chemical hazards. There is a largely open (but not unlimited) field for the use of health claims on food labels as a marketing device, and an absence of regulation imposing product redesign and recycling burdens on auto producers, although producers have cooperated in a voluntary effort to recycle mercury switches as vehicles are retired from the road. In the EU, in contrast, we find mandatory rules for the authorization of dangerous chemicals, a regime limiting health claims according to the nutrition profiles of food products, and a recycling regime for ELVs that relies on the producer-pays principle in order to create strong incentives for manufacturers to take account of dismantling, reuse, and recycling requirements in product design. Understanding these differences in regulatory outcomes requires that we examine the context in which regulatory policy making takes place in each instance: the dilemma facing regulators in the United States, in contrast with the regulatory trilemma at the core of regulatory policy making in the EU. Each of these trade-offs presents a particular set of dynamics, with different equilibrium solutions.

## Resolving the Regulatory Trilemma

Governments facing classic regulatory dilemmas pursue regulatory ends, such as protecting the environment and public health, while seeking to minimize the cost burden associated with advancing these objectives. The equilibrium balance between regulatory objectives and burdens imposed on industry will depend upon the respective mobilization capacities of environmental and industry associations as well as the resources available to the regulatory agency. Weighing the respective logics of collective action for environmentalists and producers, along with the resource constraints with which regulatory agencies such as the Environmental Protection Agency (EPA) operate,[6] voluntary industry programs that modestly advance environmental objectives, impose minimal cost burdens on industry, and require the dedication of few enforcement resources from the regulator are likely to produce equilibrium outcomes.[7] We have seen the starkest evidence of severe resource constraints in the case of the Toxic Substances Control Act (TSCA): the minimal toxic chemicals data available to the EPA; its insufficient ability to screen chemicals in the limited timeframe available; and the near impossibility of marshaling sufficient resources to meet the evidentiary standards to require additional chemicals testing—much less for more far-reaching regulatory action—imposed by courts.

As noted in chapter 2 in the discussion of regulatory preemption by industry in the US chemicals sector, voluntary initiatives are likely to demobilize environmentalists in direct proportion to the collective action problems they face; where these are high and industry faces low coordination costs—certainly the case, for example, for US auto producers, where the market is organized on oligopolistic lines—we would expect only modest voluntary programs, such as the National Vehicle Mercury Switch Removal Program created by the End of Life Vehicle Solutions Corporation.[8] Additional preemptive efforts by industry are likely to be forthcoming in proportion to the threat of mandatory regulation that emerges from new evidence of harm to the environment or human health or additional resources garnered by the regulator.[9] Withdrawal of companies from the Smart Choices food labeling program in the face of credible threats of legal action by state attorneys general and (perhaps less credible) rulemaking by the EPA substantiates the point. Varying configurations of regulatory preemption through voluntary agreements often represent equilibrium solutions to the classic regulatory dilemma in the United States.

The EU confronts a regulatory trilemma in which ends include protecting environment and public health, safeguarding industrial competitiveness, and advancing European integration (see figure 1.1). How does introduction of the integration objective into the trade-off matrix alter outcomes? First, the goal of integration elevates the tension between regulatory objectives and industrial competitiveness by lessening the appeal of voluntary regulation as an alternative. Reinforcing the above argument concerning equilibrium outcomes under the regulatory dilemma, Kathleen Segerson and Thomas J. Miceli assert that a voluntary agreement is an equilibrium outcome because of the cost savings to the regulator (1998: 111). But I argue that adding the European integration objective to the trade-off matrix disturbs this equilibrium solution. While voluntary agreements are not wholly and necessarily incompatible with integration gains, integration is most readily strengthened through compulsory regulation at the European level.

But how might mandatory EU-level regulation proceed without directly threatening industrial competitiveness? The answer is noncompliance, which provides industry time to adjust to the new regulatory environment. For some areas of regulation, much of the gain for integration comes through process—the aggregation and articulation of interests at the European level and the central role of EU institutions. In other words, mandatory regulation with a high level of noncompliance may represent an equilibrium outcome to the regulatory trilemma. This solution to the regulatory trilemma involving functional noncompliance—a configuration favoring competitiveness of industry and integration at the expense of environmental and health objectives—is reflected in the movement from A to B depicted in figure 5.1.

The application of "producer pays" to ELVs recycling provides a telling example. The European Commission temporarily resolved its regulatory trilemma by tolerating a high level of noncompliance with the 2000 ELVs directive, permitting producers time to adjust to the new regime. Given the asymmetric regulatory burden faced by auto manufacturers within the EU, the 2000 ELV directive appears to resolve the trilemma by advancing integration and environmental objectives at the expense of industrial competitiveness. However, the impact on the ability of EU auto producers to compete remains limited as long as undercompliance is widespread. In fact, all member state governments failed to transpose the directive into national law by the specified

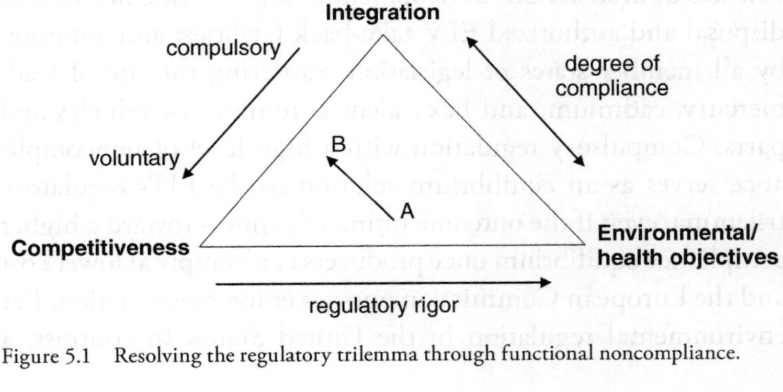


Figure 5.1 Resolving the regulatory trilemma through functional noncompliance.

deadline, requiring the Commission to begin legal action against each of them—though these proceedings stretched out over a few years.[10] Even with progress in implementation and gradual development of systems for transfer of vehicles to authorized treatment facilities and free take-back of vehicles from final owners, recent studies indicate that there is significant leakage from the recycling system through several means, from illegal export (typically to eastern Europe) of wrecked or ELV vehicles for parts to scrapping by unlicensed operators (who profit from the scrap metal but illegally dispose of the remaining hazardous materials) and abandonment of vehicles that have reached the end of their useful lives (European Parliament, 2007). Furthermore, few member states meet or are close to meeting legislated recycling and recovery targets (Commission, 2007).

While the European Commission's Environment Directorate General (DG) has an overarching interest in securing compliance with EU environmental regulations, EU institutions already have garnered many of the gains for European integration from European-level ELV recycling standards. (The arrow on the right side of figure 5.1 points in both directions, since integration may be served by compliance under some conditions and compliance deficits under others.) These include advancement of both integration and environmental objectives, such as the strengthening of coordination by producer and environmental lobbies at the European level; mutual recognition of certificates of destruction; development of an EU-wide network of disposal and authorized ELV take-back facilities; and adoption by all member states of legislation restricting the use of lead, mercury, cadmium, and hexavalent chromium in vehicles and parts. Compulsory regulation with a high level of noncompliance serves as an equilibrium solution to the EU's regulatory trilemma, even if the outcome ultimately moves toward a higher compliance equilibrium once producers can comply at lower cost and the European Commission pursues enforcement action. For environmental regulation in the United States, in contrast, a demanding regulatory regime with high noncompliance would be a stable equilibrium only under very limited conditions.

A second dynamic that differs most critically between the US context and that shaped by the EU's regulatory trilemma is the role of the EPA relative to EU institutions responsible for environmental policy. Whereas the European Commission is committed to advancing integration through European-level regulation, the EPA strives for balance between regulatory objectives and demands on agency resources; as indicated in chapter 4, the EPA seeks to retain its reputation for efficiency and to sustain its ability to reach voluntary agreements with industry in order to safeguard scarce agency resources. The EPA does not have a metaobjective equivalent to European integration. The European Commission, meanwhile, faces a powerful incentive to use the diversity of national regulatory constellations as justification for efforts to harmonize the regulatory environment across the single European market. Both organized environmental interests and industry associations typically support such initiatives, though each mobilizes to advocate their preferred approach to Europeanized regulation.

The resulting interest mobilization will produce a push and tug between environmental (or health and consumer protection) and industry positions. The outcome of this contest is indeterminate. Some institutional nodes, such as the Commission's Environment DG; its DG for Health and Consumers; and the EP's Committee on the Environment, Public Health and Food Safety, typically have been inclined—though not uniformly—to favor environmental interests. Other institutional actors, such as DG Enterprise and the EP's Internal Market Committee, are more committed to the preferences of industry. These institutional tensions provide for two possible equilibrium outcomes. One entails accommodation of industry interests within the context of a move toward compulsory EU-level regulation. This step toward "regulatory accommodation" would resemble figure 5.1, with process gains for integration resulting from European-level interest aggregation, and competitiveness protected—but by a reduction in regulatory rigor rather than systematic noncompliance. An alternative outcome involves rigorous protection of environment or health, advancing integration and these regulatory objectives while placing a heavier regulatory burden on industry, as suggested by the shift from A to C in figure 5.2.

The debate over the EU's 2006 Health and Nutrition Claims Regulation is an instance of the equilibrium represented by point C. As this example demonstrates, the motives of national government ministers meeting in the Council of Ministers may coincide with the Commission's desire to advance integration when they perceive EU regulation as a means of addressing national environmental or public health challenges. As discussed in chapter 3, national health ministers were vital to the retention of the nutrition-profiles provision of the health claims regulation. Similarly, as presented in chapter 4, national environment ministers insisted on adherence to the producer-pays principle in European regulation of ELVs recycling, and by doing so contributed to a shift in favor of that position in the EP as well.

Studies of regulatory federalism indicate that pressures from uneven standards will intensify demands for federal regulation and that these standards are likely to be high (Kelemen, 2000: 140).

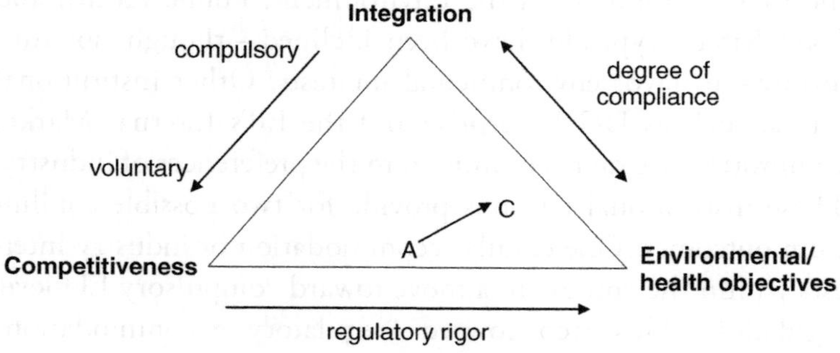


Figure 5.2 The regulatory trilemma: Regulatory rigor equilibrium.

As discussed above, I find that uneven standards are more likely to foster "federal" regulation in the EU than the United States, as in the case of ELVs. But even in the EU, the inclination of the European level to impose high standards depends on the relationship of the regulation to single-market construction and the nature of the integration benefits produced by the regulatory process. My argument coincides with the claim of the regulatory federalism literature that the existence of a patchwork of standards unleashes a dynamic likely to produce a new regulatory equilibrium (Kelemen, 2000: 141); I add that this equilibrium may include functional noncompliance. This point reinforces Dan Kelemen's argument about the expansion of EU environmental policy generally; policy implementation in response to the expansion of EU environmental policy competence in the 1970s and the 1980s was poor, but a lax approach toward poor implementation was functional: "Anticipating that they would not be forced to comply made it easier for recalcitrant Member States to accept new Community environmental regulations" (Kelemen, 2000: 151). In the post-Maastricht environment, the Commission became more aggressive about addressing implementation failures (Kelemen, 2000: 152). I argue that functional noncompliance remains an important element of environmental regulation in the EU.

## Implications for Future Regulatory Trajectories

Do these respective dynamics produced by the regulatory dilemma facing US policy makers and the regulatory trilemma with which the EU operates indicate that environmental and consumer health regulation will continue to diverge across the two systems? As I explain below, while regulatory dynamics in the United States are stable, the ascendance of the EU's focus on reduction of regulatory cost burdens may produce a declining rate of divergence of regulatory trajectories.

In the United States, we can expect to continue to see numerous forms of voluntary regulation, including preemptive self-regulation schemes offered by industry and industry-EPA agreements. These equilibrium arrangements, typically with low levels of harm abatement, will at times face perturbations that will produce higher abatement outcomes. A vivid emergent example of a potential shift to a higher environmental damage abatement equilibrium resulting from intensified mobilization of environmental interests and the threat of regulatory action comes from the increasingly controversial practice of hydraulic fracturing.[11] While the EPA in 2005 exempted hydraulic fracturing from regulation under the Safe Drinking Water Act (SDWA), it excluded the injection of diesel fuel in hydraulic fracturing fluids from this exemption. In 2003, the EPA entered into a voluntary agreement with fracturing companies B. J. Services Co. and Halliburton, in which the companies agreed to eliminate diesel fuel in hydraulic fracturing fluids injected into coalbed methane production wells in order to prevent the danger of harmful chemicals seeping into underground sources of drinking water.[12] However, by 2010, mobilization by environmental groups brought the attention of Congress to the continued threat to groundwater supplies. Congress in turn requested that the EPA conduct a study of hydraulic fracturing, scheduled to take place during 2011–2.

In August 2010, a broad coalition of environmental interest associations submitted a letter to the EPA citing evidence that B. J. Services and Halliburton had continued to inject diesel in their hydraulic fracturing operations from 2005 to 2007. In response, the EPA requested that nine companies engaged in hydraulic fracturing provide information on the chemical composition of fluids used in their fracturing processes, as well as data on their practices and well locations.[13] As an initial indication of a potential shift to a higher environmental protection equilibrium in response to congressional attention and the threat of EPA regulatory action to enforce the 2005 SDWA, the companies publicly expressed their support for an EPA study.[14] However, the regulatory threat intensified in early 2011 when several members of Congress charged in a letter to the EPA that numerous companies had violated the SDWA by using diesel in their fracturing injection fluids without seeking permits from the EPA to do so.[15] House members followed their complaint to the FDA by introducing federal legislation requiring disclosure of chemicals used in fracturing fluids.[16] As an effort to preempt federal regulation, energy companies announced the creation of a website on which they would disclose chemicals used at their production wells.[17] Even if the proposed federal regulation dies before becoming law, the resulting preemptive action by industry will represent a higher environmental and public health damage abatement equilibrium, a characteristic outcome of regulatory dilemma dynamics in the United States.

In the EU, meanwhile, the advance of the "better regulation" agenda, focused on reducing regulatory cost burdens, has heightened the prospect of approaches to the regulatory trilemma that move toward enhanced industry competitiveness at the expense of environmental objectives, with few if any integration gains. This possibility is depicted in figure 5.3 and the shift from A to D.

The movement from A to D corresponds to instances in which regulation reduces the level of environmental or health

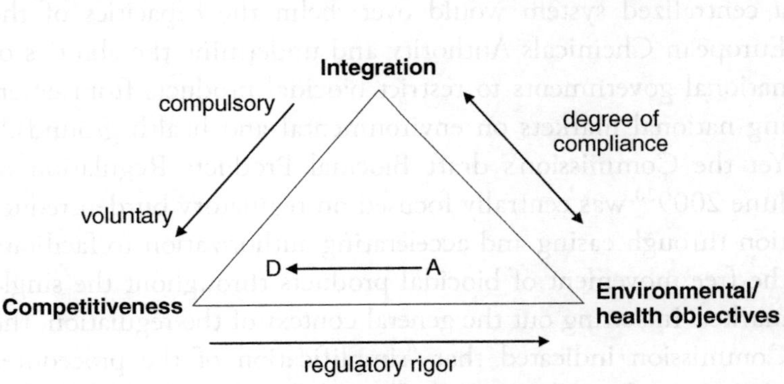


Figure 5.3 Resolving the regulatory trilemma through "better" regulation.

protection in order to meet industry needs. While the path from A to D is simplified for illustrative purposes, the revised EU regulation governing the use of biocidal products involves both environmental and consumer health issues and potentially approximates such a case. In 2008–9, the Stoiber Group—the committee of high-level experts convened to aggregate suggestions from firms, industry associations, and public authorities for reducing administrative cost of regulations—gathered proposals for reducing the burden imposed on industry by a series of environmental regulations, including the EU's 1998 Biocides Directive. The 1998 directive was motivated by the need to establish a harmonized framework for authorization and marketing of biocides—products such as household disinfectants, preservatives, and insecticides and rodenticides—and was in this sense a measure that modestly advanced both environmental objectives and integration. Proposals for regulatory revision gathered by the Stoiber Group came from the European Chemical Industry Council; the German chemicals industry federation; the European Association of Craft, Small and Medium Enterprises; and individual firms.[18] The suggestions focused on creation of a centralized European-level authorization system for all biocidal products.

In its initial response to the suggestions forwarded by the Stoiber Group, the European Commission indicated that such a centralized system would overwhelm the capacities of the European Chemicals Authority and undermine the abilities of national governments to restrict biocidal products from entering national markets on environmental and health grounds.[19] Yet the Commission's draft Biocidal Products Regulation of June 2009[20] was centrally focused on regulatory burden reduction through easing and accelerating authorization to facilitate the free movement of biocidal products throughout the single market. In setting out the general context of the regulation, the Commission indicated that "simplification of the procedures concerning the authorisation of biocidal products in the Member

States may be beneficial in reducing costs and administrative burden for companies and public authorities alike."[21]

The 1998 directive created a list of active substances permitted in biocides, with governments able to approve products with permitted chemicals under specified conditions. The 2009 proposal from the Commission provided for an EU-wide authorization system. Environmental groups[22] criticized the proposal for failing to require replacement of the most hazardous biocides with suspected carcinogenic, immunotoxic, or hormone disrupting properties, allowing for continued use of these products if they offer net benefits and can not easily be replaced.[23] The EP and the Environment Council amended the proposal to modestly strengthen the precautionary dimension of the regulation and extend authorization and labeling requirements to manufactured articles treated with biocides, such as clothing items and furniture—in this sense, the move from A to D depicted in figure 5.3 is perhaps exaggerated for this piece of legislation. However, following political agreement of the Environment Council in December 2010, the thrust of the regulation remains to achieve "the possibility of authorising biocidal products at EU level so as to reduce the administrative burden on producers."[24]

Regulatory simplification appears to be the most formative determinant of the contours of the biocides regulation, which is likely to be finalized following completion of the EP's second reading in early 2012. The focus on reducing regulatory burdens, fostered by the institutional dynamics and industry mobilization discussed here and in chapter 1, alters outcomes of the regulatory trilemma by imposing on the European Commission, in effect, a regulatory budget constraint akin to the limits facing the US EPA due to resource scarcity. Such a constraint raises serious questions about the extent to which the Commission will have the autonomy to trade off environmental and health objectives for integration gains as it produces regulatory proposals. To the degree that the "better regulation" agenda involves a progressive shrinking of this autonomy, we may see more clearly the implications of the similar market and legal logics at work in the United States and the EU: a declining rate of divergence of trajectories of regulation protecting the health of consumers and the environment.

# NOTES

## 1 DIVERGENT REGULATORY TRAJECTORIES

1. Theofanis Christoforou suggests that the stringency of EU environmental regulation began to transcend that in the United States beginning in 1989 with the EU's ban on animal growth hormone, followed by a 1990 moratorium on the use of a hormone to increase milk production and a wide range of additional measures throughout the 1990s (Christoforou, 2004: 25).
2. Bouwen (2002; 2004) argues that the relationship between the European Parliament and both national- and European-level business associations is one of interdependence, since the EP depends upon resources provided by organized business interests in exchange for institutional access.
3. See endnote 9, below.
4. Case 178/84 *Commission v. Germany* [1987] ECR 01227.
5. Case C-221/00, *Commission of the European Communities v. Republic of Austria*; *Douwe Egberts NV v. Westrom Pharma NV and others*, Case C-239/02 [2004] ECR I-7007.
6. See Commission of the European Communities, "Proposal for a Regulation of the European Parliament and of the Council on nutrition and health claims made on foods," COM(2003) 424 final, Brussels, July 16, 2003. Retrieved at http://ec.europa.eu/food/fs/fl/fl07_en.pdf on August 24, 2010.
7. For example, O'Toole et al. (1997) find that the impact of voluntary, information-based programs designed to reduce releases of toxic chemicals is a product of the manner in which information is conveyed and the extent to which there is field-level and context-specific communication. Atlas (2007), in contrast, finds that there is little public awareness or use of Toxic Release Inventory

data and that scattered anecdotal evidence of instances of collective action in response to information available in the TRI must be weighed against the tens of thousands of facilities subject to TRI-reporting requirements.

8. For additional studies on the relative effectiveness of voluntary environmental programs (VEPs), see the two-part symposium in *The Policy Studies Journal* (Vol. 35, No. 4 and Vol. 36, No. 1), beginning with the introductory comment by Peter deLeon and Jorge E. Rivera, "Voluntary Environmental Programs: A Symposium," *The Policy Studies Journal*, Vol. 35, No. 4 (2007): 685–688. These authors indicate that voluntary programs "tend to attract 'dirtier' firms that after participation do not appear to show higher environmental performance than non-VEP members" (2007: 686). Also see Strasser (2008).
9. At the same time, it is indeed the case that as the EU's "better regulation" agenda (discussed below) has advanced, industry has with increasing intensity pressed for greater reliance on voluntary regulation. A recent example concerns the 2009 voluntary agreement on mobile phone chargers, which involved a Memorandum of Understanding signed by European mobile phone companies establishing a voluntary agreement to produce a standard phone charger as a means of reducing electronic waste. See RAPID press release IP/09/1049 of June 26, 2009; accessed at http://europa.eu/rapid/pressReleasesAction.do?reference=IP/09/1049&format=HTML&aged=0&language=EN&guiLanguage=en on December 31, 2010. Also see the Commission's Enterprise and Industry DG site at http://ec.europa.eu/enterprise/newsroom /cf /itemlongdetail.cfm?&item_id=3241&lang=en.
10. Chemicals are regulated in the United States by the 1976 Toxic Substances Control Act (TSCA). However, as I discuss in chapter 2, chemicals in commerce prior to 1979 are exempt from TSCA reporting requirements, the FDA has been extremely reserved in using its limited regulatory authority under TSCA, and the regulation is widely considered ineffective. See, inter alia, Robert V. Percival, "Environmental Legislation and the Problem of Collective Action," *Duke Environmental Law & Policy Forum*, Vol. 9 (Fall 1998): 9–27; and Cynthia Ruggerio, "Referral

of Toxic Chemical Regulation Under the Toxic Substances Control Act: EPA's Administrative Dumping Ground," *Boston College Environmental Affairs Law Review*, Vol. 17 (1989–1990): 75–122.

11. For details, see the Free Trade Agreements website of the European Commission's Enterprise and Industry Directorate General at http://ec.europa.eu/enterprise/policies/international/facilitating-trade/free-trade/index_en.htm.
12. See United States Court of Appeals for the District of Columbia Circuit, No. 98-5043, Durk Pearson and Sandy Shaw, American Preventive Medical Association and Citizens for Health, *Appellants v. Donna E. Shalala*, Secretary, United States Department of Health and Human Services, et al., Appellees, on the Bureau of National Affairs website at: http://lw.bna.com/lw/19990202/985043.htm
13. Case C-221/00, *Commission of the European Communities v. Republic of Austria*.
14. Case C-239/02.
15. I elaborate on the role of private litigants in the expansion of public sector liberalization in Mitchell P. Smith, *States of Liberalization* (Albany: SUNY Press, 2005).
16. See judgment of 10.3.1983– case 172/82 on the Eur-Lex website;accessedathttp://eur-lex.europa.eu/LexUriServ/LexUriServ.do?uri=CELEX:61982J0172:EN:PDF on March 6, 2011, especially pp. 566–567. In response to the French government's claim that their legislation was required to protect the environment, the ECJ wrote: "Clearly, the environment is protected just as effectively when the oils are sold to an authorized disposal or regenerating undertaking of another Member State as when they are disposed of in the Member State of origin" (paragraph 14, p. 566).
17. Council of the European Union, Press Release, 2490th Council Meeting, Competitiveness (Internal Market, Industry and Research), Brussels, March 3, 2003.
18. Council of the European Union, Press Release, 2451st Council Meeting, Competitiveness (Internal Market, Industry and Research), Brussels, September 30, 2002, 12293/02 (Presse 283). This and all other Council documents come from the Council documents database at http://register.consilium.europa.eu/servlet

/driver?page=Result&lang=EN&ssf=DATE_DOCUMENT+DESC&fc=REGAISEN&srm=25&md=400&typ=Simple&cmsid=638&ff_TITRE=Internal+Market%2C+Industry+and+Research&ff_FT_TEXT=&ff_SOUS_COTE_MATIERE=&dd_DATE_REUNION=&srs=1, accessed on August 7, 2010.

19. See IMCO Activity Report, June 2004–May 2009 on the EP website at http://www.europarl.europa.eu/document/activities/cont/200909/20090910ATT60424/20090910ATT60424EN.pdf, accessed on August 7, 2010.
20. The emphases on reducing the burden of regulation and on early and persistent industry consultation represent significant developments precisely because they threaten to alter the very core of the EU's "regulatory governance" model. The model is predicated on the notion that the European Commission has a comparative advantage in regulation as a relatively autonomous actor at the hub of a highly specialized information network that can address market externalities and pass along the costs of regulatory implementation to national governments and industry. On these points, see Giandomenico Majone, "The Rise of the Regulatory State in Europe," *West European Politics*, Vol. 17, No. 3 (September 1994): 77–101; and Francis McGowan and Helen Wallace, "Towards a European Regulatory State," *Journal of European Public Policy*, Vol. 3, No. 4 (December 1996): 560–576.
21. See the High Level Group's web page on the European Commission Industry and Enterprise website at http://ec.europa.eu/enterprise/policies/better-regulation/administrative-burdens/high-level-group/index_en.htm.
22. Commission working document COM(2006) 691: "Measuring administrative costs and reducing administrative burdens in the European Union."
23. Also see Competitiveness Council, Key Issues Paper submitted to the spring 2008 European Council, Brussels, February 26, 2008, document 6933/08. On early consultation, see Press Release, 2665th Council Meeting, Competitiveness, Luxembourg, June 6–7, 2005.
24. See accounts of the Competitiveness Council meetings of March 3, 2003, and November 10, 2003.

## 2 Chemicals Regulation: Courts Rule in the United States; Industry Ascendant in the EU

1. Not all shared the early optimism expressed by Gaynor. Reynolds (1977: 95), for example, also writing in the immediate aftermath of the passage of TSCA, pointed out that "it is directed solely to the amorphous, litigable, and more easily circumvented concept of 'unreasonable risk.'"
2. For example, Robert B. Haemer asserts that "[a]t the time of its enactment, TSCA held great promise as a way to control exposure to toxic substances that posed threats to health and safety." In his judgment, "That promise remains unrealized" (Haemer, 1999: 104).
3. See the "Chemical Industry Archives" website of the Environmental Working Group, at http://www.chemicalindustryarchives.org/factfiction/testing.asp.
4. According to the GAO (1994: 6), the EPA had reviewed only 2 percent of these chemicals (1,200) as of 1994.
5. These five are polychlorinated biphenyls (PCBs), chlorofluorocarbons, dioxin, asbestos, and hexavalent chromium. See Lowell Center (2003: 3).
6. This standard was set in the 1988 case, *Chemical Manufacturers Association v. US Environmental Protection Agency*, decided by the US Court of Appeals, District of Columbia Circuit. The bar for EPA rulesetting established by the court, while set high, was tempered by the court's ruling that the "unreasonable risk" standard "need not be established to a more-probable-than-not degree." The court's reasoning was that "Congress intended to authorize testing where the existence of an 'unreasonable risk' could not yet be 'reasonably predicted.' The Agency's determination that it is empowered to act where the existence of an 'unreasonable risk' cannot yet be said to be more probable than not is entirely consistent with that expression of intent" (*859 F2d 977 Chemical Manufacturers Association v. US Environmental Protection Agency*, paragraph 36; accessed at OpenJurist.org; http://openjurist.org/859/f2d/977/chemical-manufacturers-association-v-us-environmental-protection-agency).

7. "EPA's TSCA Audit," *Chemical Week*, Vol. 159, No. 18 (May 7, 1997): 42.
8. Lettie M. Wenner, "Environmental Policy in the Courts," in Norman J. Vig and Michael E. Kraft, eds., *Environmental Policy in the 1990s*, 2nd ed. (Washington, DC: CQ Press, 1994), p. 155.
9. Haemer (1999: 118) reaches a similar conclusion based on the 1990 *Corrosion Proof Fittings v. EPA asbestos* case: "The fact that the court found ten years of rulemaking and a 45,000 page record inadequate to support a ban on asbestos makes it appear that EPA management has good reason to avoid rulemaking altogether."
10. There were a much larger number of Clean Air Act cases than TSCA cases over the 1995–2002 period studied by Kamieniecki; the EPA was the winning party in 14 of 34 Clean Air Act cases, with split decision in 9 of these cases; while for TSCA, the EPA was defeated in 3 of 4 cases. See Kamieniecki, 2006: 139.
11. As Joel Reynolds points out, for example, Dow Chemical compiled and presented before hearings of the House Subcommittee on Consumer Protection and Finance of the House Committee on Interstate and Foreign Commerce, an extensive list of existing regulatory measures as evidence in support of its claim that additional regulation of toxic substances was unnecessary (Reynolds, 1977: 38, fn 10).
12. Representatives of the Manufacturing Chemists Association argued during hearings that eventually led to TSCA that "regulation would halt the march of chemical progress"; in its 1971 congressional testimony, the association contended that "any law of this nature is bound to impede innovation to some degree. In the new product area, compliance with a law of even selective scope will be expensive, time consuming and detrimental to the introduction of new chemical products." Environmental Working Group (2009b).
13. Using data from the Bureau of Economic Analysis of the Environmental Economics Division of the US Department of Commerce from the mid-1990s, Palmer, Oates, and Portney cite industry pollution control spending in excess of $100 billion; their informal survey data on efficiency "offsets" suggest these amount to less than 2 percent of regulatory costs (Palmer, Oates, and

Portney, 1995: 127–128). This would create significant incentives for firms to invest in preempting more rigorous controls.

14. For an account of the United Kingdom's "Responsible Care" program, which followed the US program by mere months, see Simmons and Wynne, 1993. The program of the United Kingdom's Chemical Industries Association is similar to the US version in that the industry association requires participation by all member firms and is not directly involved in enforcement; the program does not provide for external evaluation; and the program is primarily motivated by rising public mistrust of the sector.
15. See Gary E. R. Hook, "Responsible Care and Credibility," *Environmental Health Perspectives*, Vol. 104, No. 11 (November 1996): 1138–1139.
16. According to the Environmental Working Group (2009a), the CMA spent between $1 million and $2 million annually to implement Responsible Care at member companies during the 1990s, but in excess of $10 million publicizing the program.
17. See "Ready to Play Offense?" *Chemical Week*, Vol. 161, No. 45 (December 1, 1999): 5.
18. The CMA did eventually introduce a system of "third-party verification" in the form of mandatory posting of information such as workplace-safety data on a public website.
19. According to the Environmental Working Group, "Nothing in Responsible Care commits any company or facility to measurable goals for reducing chemical hazards, to timelines to meet such goals, or to objective assessment of progress by independent outside authorities" (2009a: 2).
20. At the same time, Rees suggests that Responsible Care has generated an "ethical framework" and norms of accountability that may lead to enhanced efforts by firms to meet commitments to reduce toxic wastes and engage in good product stewardship (1997: 512).
21. The first evidence of the danger of polybrominated diphenyl ethers (PDBEs) was uncovered in a study of levels in breast milk in Sweden from 1972 to 1997 conducted by a research team at Stockholm University. The findings prompted a ban of these substances by a directive of the EU's Council of Ministers and the EP.

22. See speech by John Hontelez, Secretary General, European Environmental Bureau, European Voice Conference, "Beyond REACH," April 1, 2003, Brussels. Available at: www.eeb.org/activities/chemicals/ 27-03-03-Speech-John-Hontelez-EV-Chemicals-Conference.pdf.
23. See "What We Need from reach: Views on the Proposal for a New Chemical Legislation within the eu," The International Chemical Secretariat, January 2005; www.chemsec.org.
24. See "Toxicity Tests Stripped from reach Proposal," press release of The Greens/ European Free Alliance in the European Parliament, September 13, 2005 (http://www.greens-efa.org/en/press/detail.php?id=2690&lg=en); and Leigh Thomas, "eu Chemicals Reform Reaching Critical Point," *Agence France Presse*, September 18, 2005; accessed via LexisNexis Academic, http://web.lexis-nexis.com/universe. Also see "Background Note on the Results of the Vote on the Proposed New Chemicals Legislation (REACH)," European Parliament, Committee on the Environment, Public Health and Food Safety, Strasbourg, November 17, 2005.
25. The Competitiveness Council's remit is "systematic and comprehensive impact assessment of proposed Community legislation"; ensuring that the Commission consults and takes account of the concerns of the business community in the policy-making process; and rebalancing the Lisbon agenda to take account of the objectives of competitiveness and growth. Council of the European Union, Press Release, 2490th Council Meeting, Competitiveness (Internal Market, Industry and Research), Brussels, March 3, 2003. Obtained from the Council documents database at: http://register.consilium.europa.eu/servlet/driver?page=Result&lang=EN&ssf=DATE_DOCUMENT+DESC&fc=REGAISEN&srm=25&md=400&typ=Simple&cmsid=638&ff_TITRE=Internal+Market%2C+Industry+and+Research&ff_FT_TEXT=&ff_SOUS_COTE_MATIERE=&dd_DATE_REUNION=&srs=1.
26. At its November 2003 meeting, the Competitiveness Council cited the conclusions of the October 2003 European Council (of heads of government) on competitiveness, which made specific reference to the case of REACH: "European Union legislation should not be a handicap to EU competitiveness compared to that of other major economic areas. To this end the Commission is

invited to take into account the consequences of proposed EU legislation on enterprises through providing a comprehensive impact assessment. The forthcoming proposal on chemicals, which will be examined by the Competitiveness Council in coordination with other Council configurations, will be the first case for implementing this approach." Press Release, 2539th Council meeting, Competitiveness (Internal Market, Industry and Research), Brussels, November 10, 2003, p. 10.

27. See accounts of the Competitiveness Council meetings of March 3, 2003, and November 10, 2003.
28. Press release, 2583rd Council Meeting, Competitiveness, Brussels, May 17–18, 2004, p. 9, paragraph 17; retrieved from http://www.consilium.europa.eu/uedocs/cms_Data/docs/pressdata/en/intm/80522.pdf.
29. According to Corporate Europe Observatory (2005: 2), CEFIC employs a staff of 140 in Brussels and is comprised of large chemical companies as well as national chemicals sector industry associations.
30. In January 2007, UNICE became "BusinessEurope."
31. See letter on "Impact of REACH on downstream users" to the Competitiveness Council of the EU of November 23, 2006, European Committee of Domestic Equipment Manufacturers; http://www.ceced.org/.
32. See EEB and WWF, "Business Impact Assessments and the Work by KPMG for UNICE and CEFIC." Available at: http://www.eeb.org/activities/chemicals/20050113-EEB-WWF-KPMG-brief-final.pdf.
33. See "Note on the Studies Undertaken in the Framework on the Memorandum of Understanding on Further Work Concerning the Impact Assessment of REACH," April 27, 2005; accessed at http://ec.europa.eu/environment/chemicals/reach/background/docs/memorandum_of_understanding.pdf.
34. See "EU Officials Seek to Make Sense of 36 REACH Reports," November 3, 2004. The press release issued by the Dutch presidency on October 27, 2004, concluded that "[g]overnment officials, whether they belonged to the environment, health or economic affairs ministries, agreed with the objectives of the legislation, but judged the costs for business to be

too high. They concluded that the European Commission and the governments of the EU member states must do all they can to limit the negative effects for business while maintaining the benefits. The costs will mainly affect business, especially Small and Medium-Sized Enterprises (SMEs). Moreover, REACH may affect the competitive position of European companies on the global market, if the direct and indirect costs are not kept under control"; both documents accessed at http://www.euractiv.com/Article?tcmuri=tcm:29-131725-16&type=News.

35. The Internal Market and Consumer Affairs Committee is the newest EP committee, created in 2004. IMCO's brief is the internal market, competitiveness, and economic protection of consumers (protection of consumer health falling under the remit of ENVI).
36. Rules of Procedure of the European Parliament, 16th edition, July 2004.
37. The rules state that "the committee responsible shall accept without a vote amendments from the committee asked for an opinion where they concern matters which the chairman of the committee responsible considers, on the basis of Annex VI, after consulting the chairman of the committee asked for an opinion, to fall under the competence of the committee asked for an opinion, and which do not contradict other elements of the report." Rules of Procedure of the European Parliament, 16th ed., July 2004, article 47; accessed at http://www.europarl.eu.int/omk/sipade3?SAME_LEVEL=1&LEVEL=2&NAV=X&DETAIL=&PUBREF=-//EP//TEXT+RULES-EP+20040720+TOC+DOC+XML+V0//EN.
38. See "The New Chemicals Policy (REACH): An mep's Views on the Upcoming Parliamentary Debate," speech of MEP and Industry Committee Rapporteur for REACH Elly Plooij-van Gorsel to the Conference of the Industrial Minerals Association—Europe Conference, May 13, 2004; http://www.plooij.nl/engels/toespraken/096.htm.
39. See Secretariat of the Committee on the Environment, Public Health and Food Safety, European Parliament, "Background Note on the Results of the Vote on the Proposed New Chemicals Legislation (REACH)," Strasbourg, November 17, 2005; and "REACH: MEPs Strike a Balance between Health, the

Environment and Industry," European Parliament news headlines, http://www.europarl.eu.int/news/public/story_page/008-2560-318-11-46-901-20051118STO02559-2005-14-11-2005/default_en.htm; downloaded on February 20, 2006.

40. E-mail correspondence, January 16, 2006.
41. Interview with EU Policy Officer, European Environmental Bureau, April 3, 2006; interview with Chemicals Campaigner, Friends of the Earth Europe (FOEE), April 5, 2006.
42. E-mail correspondence with an official of EEB, January 23, 2006.
43. Calculated from EP (2005).
44. European Report, No. 2995, Oct 8, 2005; http://web.lexis-nexis.com/universe.
45. E-mail correspondence, January 17, 2006.
46. This amendment to paragraph 38 of the EP report eliminated persistent, bioaccumulative and toxic substances and endocrine disruptors from the authorization system; it was approved by a vote of 240 for and 201 against, with 20 abstentions. See the Friends of the Earth Europe website at http://foeeurope.org/euvotewatch/support_docs.
47. See Press Release, 2624th Council Meeting, Competitiveness (Internal Market, Industry and Research), Brussels, November 25–26, 2004, p. 15; and Press Release, 2681st Council Meeting, Competitiveness (Internal Market, Industry, and Research), Luxembourg, October 11, 2005, p. 9.
48. Press Release, Extraordinary Council meeting, Competitiveness (Internal Market, Industry, and Research), Brussels, December 13, 2005. The Council's statement explains: "While some delegations stressed the importance of providing for strong incentives or even requirements to substitute dangerous substances, other delegations were worried by the impact on industry if excessive conditions for authorisation were to be adopted. The agreement by the Council strikes a balance between these different views."
49. For example, in May 2002, the European Commission issued a new impact assessment strategy, to be applied to "all major" policy initiatives—those in the Commission's annual "Work Programme." See "Communication from the Commission on Impact Assessment," Brussels, COM(2002) 276 final.

50. Discussion via telephone with Gregg Small, Executive Director, Washington Toxics Coalition, May 1, 2006.
51. See the Louisville Charter website at http://www.louisvillecharter.org/thecharter.shtml.
52. See the Health Care Without Harm website at http://www.noharm.org/us/.
53. The full text of the legislative proposal is available on the US Congress tracking website, GovTrack, at http://www.govtrack.us/congress/billtext.xpd?bill=h109-4308.
54. See http://www.govtrack.us/congress/bill.xpd?bill=h109-4308.
55. The GAO maintains a list of programs "at high risk for waste, fraud, abuse, mismanagement or in need of broad reform," and updates the list every two years. "Transforming EPA's Process for Assessing and Controlling Toxic Chemicals" was one of 31 items on the list issued in January 2009. See the GAO's High Risk List at http://www.gao.gov/highrisk/risks/high_risk.php.
56. The GAO's explanation for including assessment and control of toxic chemicals on its high-risk list of areas requiring congressional and executive branch attention includes a critique of TSCA, which "generally places the burden of obtaining information about the roughly 80,000 chemicals already on the U.S. market on EPA, rather than on the companies that produce the chemicals." See http://www.gao.gov/highrisk/risks/safety-security/epa_and_toxic_chemicals.php.
57. See "Bill Seeks to Revamp u.s. Chemical Laws," *The Washington Post*, April 15, 2010, p. A6.
58. See Mike Verespej, "States Urge Feds to Tighten Chemical Rules," *Waste & Recycling News*, January 18, 2010, p. 9; and Kara Sissell, "Industry, NGOs Debate TSCA Reform Priorities," *Chemical Week*, April 5–12, 2010, p. 15; both accessed via Lexis Nexis Academic database. Also see Michael Virtanen, "State Working on Chemical List," *Newsday*, January 4, 2010, p. A13. This article concerns a movement by the state of New York to avoid toxics in its purchases and reports that Massachusetts, California, Maine, and Washington also have adopted this practice.
59. In August 2009, the American Chemistry Council announced its "10 Principles for Modernizing TSCA," in response to the congressional hearing process. See http://www.americanchemistry.com/s_acc/sec_article_acc.asp?CID=2178&DID=9939.

## 3 Health Claims on Food Labels: Protecting Free Commercial Speech versus Completing the Single Market

1. "Guidance for Industry: Letter Regarding Point of Purchase Food Labeling," US Food and Drug Administration; accessed on September 20, 2010, at http://www.fda.gov/Food/GuidanceComplianceRegulatoryInformation/GuidanceDocuments/FoodLabelingNutrition/ucm187208.htm.
2. As Blim (1994: 738) describes, there was an attempt in 1976 by a bread manufacturer to make a claim regarding the beneficial effects of the fiber contained in the product, and the FDA responded by calling for the company either to delete the claim or submit an application for approval of the product as a new drug.
3. The food industry sought to expand this protection from regulatory diversity by calling for extension of the NLEA preemption clause to health warnings regarding allergens and toxins, as well as labels concerning the "natural" or "organic" status of the food product. However, labels addressing these issues remained outside the scope of the NLEA (Bradley, 1994: 659–660).
4. These related to calcium and osteoporosis; fat and cancer; saturated fat, cholesterol and coronary heart disease; fiber-containing grain products, fruits, and vegetables and cancer; fruits, vegetables, and grain products that contain fiber and risk of coronary heart disease; sodium and hypertension; and fruits and vegetables and cancer. See "Guide to Nutrition Labeling and Education Act (NLEA) Requirements," US Food and Drug Administration, Office of Regulatory Affairs; available at http://www.fda.gov/ora/inspect_ref/igs/nleatxt.html#HEALTH%20CLAIMS.
5. The four proposed claims concerned a link between antioxidant vitamins and reduced risk of certain kinds of cancers; consumption of fiber and the risk of colorectal cancer; omega-3 fatty acids and reduced risk of coronary heart disease; and the superior effectiveness in reducing the risk of neural tube defects of folic acid in dietary supplements relative to lower amounts in foods in common form. See the decision on the Bureau of National Affairs website at http://lw.bna.com/lw/19990202/985043.htm.

6. See United States Court of Appeals for the District of Columbia Circuit, No. 98-5043, Durk Pearson and Sandy Shaw, American Preventive Medical Association and Citizens for Health, *Appellants v. Donna E. Shalala*, Secretary, United States Department of Health and Human Services, et al., Appellees, on the Bureau of National Affairs website at http://lw.bna.com/lw/19990202/985043.htm.
7. *Whitaker v. Thompson* 248 F. Supp. 2d 1 (D.D.C. 2002).
8. See Memorandum Opinion, United States District Court for the District of Columbia, *Alliance for Natural Health U.S. v. Kathleen Sebelius, et al.*, Civil Action No. 09-01470 (ESH); accessed at https://ecf.dcd.uscourts.gov/cgi-bin/show_public_doc?2009cv1470-28 on December 7, 2010, pp. 12, 24.
9. Memorandum Opinion, United States District Court for the District of Columbia, *Alliance for Natural Health U.S. v. Kathleen Sebelius, et al.*, p. 39.
10. Ibid., p. 22, note 21.
11. Public docket comments on "Consumer Health Information for Better Nutrition Initiative," http://www.fda.gov/Food/LabelingNutrition/LabelClaims/QualifiedHealthClaims/ QualifiedHealthClaimsPetitions/ucm096175.htm, letter of April 10, 2003; accessed August 19, 2010.
12. Public docket comments on "Consumer Health Information for Better Nutrition Initiative," letter of May 27, 2003, from Representatives Henry Waxman, Edward Markey, David Price, and Jeff Bingaman and Senator Barbara Boxer to FDA Commissioner Mark McClellan.
13. Public docket comments on "Consumer Health Information for Better Nutrition Initiative," letter of May 16, 2003.
14. US Food and Drug Administration, Consumer Health Information for Better Nutrition Initiative: Task Force Final Report, July 10, 2003. Furthermore, in explaining why the FDA chose to extend the logic of decisions regarding dietary supplements to conventional foods, the task force report indicated that "FDA's regulations for health claims for dietary supplement labeling are identical in all material respects to the NLEA provisions for health claims for conventional food labeling. Put another way, FDA adopted the same procedure and standard for health claims

for dietary supplement labeling that Congress prescribed in the NLEA for health claims in conventional food labeling. These dietary supplement regulations, like the NLEA provisions in question, do not provide for qualified claims. Hence, based on Pearson and related cases, a court faced with a decision by FDA to not permit a qualified health claim for a conventional food might well find the same tension between the NLEA provisions and the First Amendment"; accessed on August 16, 2010, at http://www.fda.gov/Food/LabelingNutrition/LabelClaims/QualifiedHealthClaims/QualifiedHealthClaimsPetitions/ucm096010.htm

15. See the FDA's website on "Qualified Health Claims: Letters of Enforcement Discretion" at http://www.fda.gov/Food/LabelingNutrition/LabelClaims/QualifiedHealthClaims/ucm072756.htm; accessed on August 17, 2010.
16. *Alliance for Natural Health U.S. v. Kathleen Sebelius, et al.*, No. 1:09-cv-01470 (D.D.C. filed August 4, 2009).
17. See, inter alia, Joint Comments of Julian M. Whitaker, M.D., and Wellness Lifestyles, Inc., to the FDA's Task Force on Consumer Health Information for Better Nutrition, Public Docket 03N-0069; accessed at http://www.fda.gov/Food/LabelingNutrition/LabelClaims/QualifiedHealthClaims/QualifiedHealthClaimsPetitions/ucm096175.htm on December 6, 2010; and Ricardo Carvajal, "Alliance For Natural Health Takes a Swing at FDA's Regulation of Qualified Health Claims," August 13, 2009, FDA Law Blog; Hyman, Phelps & McNamara, P.C.; accessed at http://www.fdalawblog.net/fda_law_blog_hyman_phelps/2009/08/alliance-for-natural-health-takes-a-swing-at-fdas-regulation-of-qualified-health-claims.html on December 7, 2010.
18. "The basic principles of SSA articulated in the 1999 guidance have not changed. A finding of SSA still requires the agency's best judgment as to whether qualified experts would likely agree that the scientific evidence supports the substance/disease relationship that is the subject of a proposed health claim." FDA, "Guidance for Industry, Evidence-Based Review System for the Scientific Evaluation of Health Claims," Available at http://www.fda.gov/Food/GuidanceComplianceRegulatoryInformation/GuidanceDocuments/FoodLabelingNutrition/ucm073332.htm; accessed December 8, 2010.

19. US Food and Drug Administration, "Guidance for Industry: Evidence-Based Review System for the Final Evaluation of Health Claims—Final," January 2009, note 25; accessed at http://www.fda.gov/Food/GuidanceComplianceRegulatoryInformation/GuidanceDocuments/FoodLabelingNutrition/ucm073332.htm on September 20, 2010.
20. See "Proposal for a Regulation on Nutrition and Health Claims made on Foods," Briefing Note (IP/A/ENVI/OF/2005-199), DG Internal Policies of the Union, Policy Department: Economic and Scientific Policy; requested by the EP's Committee on the Environment, Public Health and Food Safety; accessed at http://www.europarl.europa.eu/activities/committees/studies/download.do?language=en&file=12883 on August 24, 2010.
21. DG SANCO Discussion Paper on Nutrition Claims and Functional Claims, NFPA, July 20, 2001, p. 3; accessed at http://ec.europa.eu/food/food/labellingnutrition/resources/fl_com9186.pdf on August 24, 2010.
22. Council Directive 79/112/EEC on the labeling, presentation, and advertising of foodstuffs. For the text, see http://eur-lex.europa.eu/LexUriServ/LexUriServ.do?uri=CELEX:31979L0112:EN:HTML; accessed on August 28, 2010.
23. Judgment of the Court (Sixth Chamber), January 23, 2003, Failure by a Member State to Fulfill Its Obligations—Approximation of laws—Articles 28 EC and 30 EC—Directive 79/112/EEC—Labelling and presentation of foodstuffs. Available at http://curia.europa.eu/jurisp/cgi-bin/form.pl?lang=en&alljur=alljur&jurcdj=jurcdj&jurtpi=jurtpi&jurtfp=jurtfp&numaff=C-221/00&nomusuel=&docnodecision=docnodecision&allcommjo=allcommjo&affint=affint&affclose=affclose&alldocrec=alldocrec&docor=docor&docav=docav&docsom=docsom&docinf=docinf&alldocnorec=alldocnorec&docnoor=docnoor&radtypeord=on&newform=newform&docj=docj&docop=docop&docnoj=docnoj&typeord=ALL&domaine=&mots=&resmax=100&Submit=Rechercher.
24. *Douwe Egberts NV v. Westrom Pharma NV and others*, Case C-239/02 [2004] ECR I-7007.
25. Council Directive 90/496/EEC on nutrition labeling for foodstuffs of September 24, 1990. *Official Journal of the European*

*Communities* L276 of October 1990. Luxembourg: EC (1990): 40–44; accessed at http://eur-lex.europa.eu/smartapi/cgi/sga_doc?smartapi!celexapi!prod!CELEXnumdoc&lg=EN&numdoc=31990L0496&model=guichett on August 25, 2010.

26. "CIAA Comments on DG SANCO Discussion Paper on Nutritional and Functional Claims," SANCO/1341/2001, p. 2; accessed on European Commission's food safety website at http://ec.europa.eu/food/food/labellingnutrition/resources/fl_com9165.pdf on August 24, 2010.
27. Swedish Food Federation to Commission discussion paper; accessed at http://ec.europa.eu/food/food/labellingnutrition/resources/fl_com9191.pdf on August 24, 2010.
28. Response to the Commission's DG SANCO (Safety and Consumer Protection) Discussion Paper on Nutrition Claims and Functional Claims, NFPA, July 20, 2001, p. 3; accessed at http://ec.europa.eu/food/food/labellingnutrition/resources/fl_com9186.pdf on August 24, 2010.
29. NFPA response to Commission Discussion Paper, p. 5.
30. NFPA response, pp. 5–6.
31. The European Food Safety Authority, located in Parma, Italy, was established in 2002 to provide independent scientific advice to the EU institutions and national governments to guide risk management and regulation in the areas of food safety and nutrition. EFSA's mandate is defined by Regulation (EC) No 178/2002, January 28, 2002.
32. See "FDF Commentary in Response to EC Discussion Paper (SANCO/1341/2001) on Nutrition Claims and Functional Claims;" accessed at European Commission's food safety website http://ec.europa.eu/food/food/labellingnutrition/resources/fl_com9176.pdf on August 24, 2010.
33. See "Euro Food and Supplement Industries Balk at Revised Health Claim Proposal," *Nutraceuticals International*, Vol. 8, No. 8 (August 2003); accessed via http://web.lexis-nexis.com.
34. See "MEPs Drop Compulsory Labeling for 'Health' Food," www.euractiv.com, April 25, 2005.
35. The quotations in this paragraph are taken from the EP website, Debates, Wednesday, May 25, 2005; available at http://www.europarl.europa.eu/omk/sipade3? accessed on October 18, 2009.

36. "Consumer Affairs: Parliament Waters Down Key Nutrition Claims Regulation," *European Report* No. 2965, May 27, 2005. Also see "Food Labelling Rules Are Watered Down," *The Irish Times*, May 27, 2005.
37. Council of the European Union, note from Council Secretariat, Working Party on foodstuffs, February 17, 2004; interinstitutional file 2003/0165 (COD); accessed at http://www.ihta.org/content/CouncilClaims17Feb.pdf on August 24, 2010.
38. See comments of rapporteur Adriana Poli Bortone, verbatim debate of the EP, May 15, 2006, Strasbourg; accessed at http://www.europarl.europa.eu/sides/getDoc.do?type=CRE&reference=20060515&secondREF=ITEM-016&format=XML&language=EN on August 25, 2010.
39. The EPP-DE had a delegation of 268 at the outset of the sixth EP, out of 732 total seats. Both numbers increased with the accession of Bulgaria and Romania to the EU in 2007.
40. Verbatim debate of the EP, May 15, 2006, Strasbourg.
41. Statements of Schnellhardt, Doyle, and Krahmer all from verbatim debate of the EP, May 15, 2006.
42. Sommer and Ulmer statements from verbatim debate of the EP, May 15, 2006.
43. Verbatim debate of the EP, May 15, 2006.
44. Corbey, Evans, and Liotard comments from verbatim debate of the EP, May 15, 2006.
45. Verbatim debate of the EP, May 15, 2006. The International Sweets and Biscuits Fair (ISM) is the world's largest confectionary trade fair and has taken place annually in Cologne since 1971. The 2011 show included more than 1,400 exhibitors from 65 countries, with more than 32,000 visitors—principally industry wholesalers—from 144 countries. See http://www.expodatabase.com/messe/ism-international-sweets-and-biscuits-fair-cologne-686.html; accessed May 7, 2011.
46. From Commission Food Safety Labelling & Nutrition—Health & Nutrition claims website, http://ec.europa.eu/food/food/labelling-nutrition/claims/health_claims_en.htm; accessed May 7, 2011.
47. Rapid press release IP/10/1176 of September 27, 2010; accessed at http://europa.eu/rapid/pressReleasesAction.do?reference=IP/10/1

176&format=HTML&aged=0&language=EN&guiLanguage=fr on May 7, 2011.

48. See http://www.euractiv.com/en/food/eu-delays-decision-food-health-claims-news-498261; published September 30, 2010—updated November 26, 2010; accessed May 7, 2011.
49. The Commission announced its proposal on January 30, 2008; the EP issued its position on the proposal on June 16, 2010, and the Council of Ministers adopted its common position on February 22, 2011. See the European Commission's food safety website at http://ec.europa.eu/food/food/labellingnutrition/foodlabelling/proposed_legislation_en.htm
50. Richard Clarke, "Nutrient Profiles Face Axe from Nutrition & Health Claims Regs," *Functional Ingredients Newsletter*, April 30, 2010, http://newhope360.com/nutrient-profiles-face-axe-nutrition-health-claims-regs; accessed May 7, 2011.
51. Richard Clarke, "MEPs Reject Bid to Delete Nutrient Profiles from Health Claims Regulation," *Functional Ingredients Newsletter*, June 21, 2010, http://newhope360.com/meps-reject-bid-delete-nutrient-profiles-health-claims-regulation; accessed May 7, 2011. The proposed amendment was controversial from the outset because it represented a legal—but unusual—attempt to modify a piece of legislation retroactively. Adding to the controversy, several MEPs announced after the vote that they had mistakenly registered the opposite position of what they had intended (five against the amendment and two in favor); if true the EP would have cut nutrition profiles in its first reading of the new Food Information legislation. See Richard Clarke, "Fate of Nutrient Profiles in Doubt after MEPs Reveal Voting Blunders," *Functional Ingredients Newsletter*, August 10, 2010, http://newhope360.com/fate-nutrient-profiles-doubt-after-meps-reveal-voting-blunders; accessed May 7, 2011.
52. "EU Food Labelling Rules Under Attack," EurActiv.com, January 18, 2011, http://www.euractiv.com/en/cap/eu-food-labelling-rules-attack-news-501333; accessed May 8, 2011.
53. Companies involved in Smart Choices paid a $100,000 fee and agreed to forego their own labeling systems. Members included Kellogg's, Kraft Foods, ConAgra Foods, Unilever, General Mills,

PepsiCo, and Tyson Foods. See William Neuman, "For Your Health, Fruit Loops," *The New York Times*, September 4, 2009, Business Section.

54. FDA letter to the Smart Choices program, August 19, 2009; accessed at http://www.fda.gov/Food/LabelingNutrition/LabelClaims/ucm180146.htm on August 17, 2010.
55. Jacobson resigned in response to what he viewed as food industry domination of the project. See Neuman, *The New York Times*, September 4, 2009.
56. FDA letter to the Smart Choices program, August 19, 2009.
57. Blumenthal's office announced the investigation on October 15, 2009, indicating that the focus would be on the use of consumer research in devising the Smart Choices program and the role major food manufacturers may have played in paying for or developing the program. See Connecticut Attorney General's Office Press Release, " Attorney General Investigates 'Smart Choices' Food Labels That Endorse Mayonnaise and Sugary Cereals," http://www.ct.gov/ag/cwp/view.asp?Q=448878&A=3673; and press release of October 29, 2009, "Attorney General Announces All Food Manufacturers to Drop Smart Choices Logo," http://www.ct.gov/ag/cwp/view.asp?A=3673&Q=449880; both accessed May 8, 2011. Also see William Neuman, "Connecticut to Scrutinize Food Labels," *The New York Times*, October 14, 2009; accessed at http://www.nytimes.com/2009/10/15/us/15food.html on May 8, 2011.

## 4 Recycling of End-of-Life Vehicles: "Good Product Stewardship" in the United States; Producer Pays in the EU

1. Directive 2002/96/EC of the European Parliament and of the Council of January 27, 2003, on WEEE; accessed on the Eur-Lex website at http://eur-lex.europa.eu/LexUriServ/LexUriServ.do?uri=ELEX:32002L0096:EN:NOT on February 11, 2010.
2. Tracing the legislative process from the Commission's proposal through Council deliberations and two readings in the EP, Golub finds that the European Commission failed to secure most of its major objectives: states advocating more flexible and less demanding recycling and recovery targets, limits on opt-outs for higher

targets by individual governments, and long derogations for less green member states, carried the day. In this regard, the waste packaging directive followed a different trajectory from the ELVs directive, in which the core elements of the proposal of the Commission's Environment Directorate General survived the process of negotiation between and within the Council of Ministers and the EP.

3. US EPA, "Product Stewardship"; accessed at www.epa.gov/epr/about/index.htm on February 12, 2007.
4. US EPA "Product Stewardship" website.
5. See Chaz Miller, "States Lead the Way: Pioneering Recycling Efforts in the U.S.," *Waste Management World* (Sep/Oct 2006): 35–38; accessed at http://www.waste-management-world.com/index/display/article-display/273718/articles/waste-management-world/volume-7/issue-5/recycling-special/states-lead-the-way-pioneering-recycling-efforts-in-the-us.html on January 10, 2011.
6. See Julie C. Becker, "A 'Cradle to Cradle' Debate in Congress," *EPA Journal* Vol. 18 (July/August 1992): 35–37.
7. On the concept of a regulatory budget constraint, see Giandomenico Majone, "The Rise of the Regulatory State in Europe," *West European Politics* Vol. 17, No. 3 (September 1994): 77–101.
8. On the proposed national computer recycling measure, see "Growth of E-Waste May Lead to National 'E-Fee' on Devices," *Computerworld* (February 28, 2007); accessed at http://www.computerworld.com/s/article/9012018/Growth_of_e_waste_may_lead_to_national_e_fee_on_devices. The Maine law states that "[i]t is the intent of the Legislature that manufacturers of electronic products and components will be responsible for ensuring proper handling, recycling and disposal of discarded products and that costs associated with consolidation, handling and recycling be internalized by the manufacturers of electronic products and components before the point of purchase." Maine E-Waste Recycling Law, revised June 2011, available at http://www.maine.gov/dep/rwm/ewaste/pdf/38mrsasec1610_2011revisions.pdf. On the California law, Danielle M. Bergner writes, "Although the Act requires manufacturers to submit annual reports, it does not require that the manufacturers actually do anything to reduce the quantity of hazardous materials or increase the use of recycled materials in their products" (Bergner, 2004: 384).

9. The "NEPSI Compromise Resolution," adopted in Portland, Oregon, in February 2004, may be found at http://www.ecy.wa.gov/programs/swfa/mrw/pdf/TechnicalPapers/NEPSI%20Final%20Resolution.pdf. On NEPSI see http://www.sustainelectronics.illinois.edu/resources/fullrecord.cfm?id=1047 and *Pollution Engineering* (April 2004): 10; both accessed via Lexis-Nexis at http://content.ebscohost.com.ezproxy.lib.ou.edu; all sources viewed on January 10, 2011.
10. For additional information about the materials in end-of-life autos and their treatment, see N. Kanari, J. L. Pineau, and S. Shallari, "End-of-Life Vehicle Recycling in the European Union," *Journal of Metallurgy* (August 2003), online version, at http://www.tms.org/pubs/journals/JOM/0308/Kanari-0308.html.
11. The Swedish government promulgated ordinances on producer responsibility for newsprint and tires in 1994 and packaging in 1997. See "Producer Responsibility in Sweden," paper of Jonas Ebbesson, Professor of Environmental Law, Stockholm University, presented at the Avosetta meeting (a group of lawyers committed to the development of environmental law in the EU and member states; see www.avosetta.org), Brussels, January 16–17, 2004; accessed at http://www-user.uni-bremen.de/~avosetta/sweden.pdf. As Ebbesson explains, these ordinances are predicated on the notion of voluntary compliance rather than strict enforcement action.
12. EPA Waste website, Automotive Parts; http://www.epa.gov/epawaste/conserve/materials/auto.htm.
13. Voluntary partnerships include a cooperative research and development agreement between a group of auto manufacturers, the US Department of Energy, and the American Plastics Council designed to address some of the recycling problems posed by ELVs. On the issue of preemptive action, see Gary A. Davis, "Automotive Take-Back and Recycling Programs," case study prepared for Center for Clean Products and Clean Technologies, University of Tennessee, section 5.2.4.; accessed at http://www.p2pays.org/ref/09/08872/eprn5-8.pdf on August 19, 2010.
14. See http://www.elvsolutions.org/.
15. See the EPA press release at http://yosemite.epa.gov/opa/admpress.nsf/1e5ab1124055f3b28525781f0042ed40/030075aab3c88984

852578d300566b3b!OpenDocument. The interagency report is available at http://www.epa.gov/wastes/conserve/materials/ecycling/taskforce/docs/strategy.pdf; both accessed on July 22, 2011.

16. The ten are Austria, Belgium, France, Germany, Italy, the Netherlands, Portugal, Spain, Sweden, and the United Kingdom, collectively representing 96 percent of the estimated ELVs in the (then) 15 member states of the EU. Only Greece and Luxembourg did not have plans to introduce an ELV regime. See Zaboli et al. (2000: I.38).
17. Michael Mann, "EU Car Recycling Law in Doubt as Industry Stalls," *Reuters News Service*, June 17, 1999.
18. See "End-of-Life Vehicle Directive Blocked Again," *Automotive Environment Analyst*, July 1, 1999; accessed via LexisNexis Academic database at http://web.lexis-nexis.com.ezproxy1.lib.ou.edu/universe/form/academic/s_guidednews.html? on February 12, 2009.
19. 2165 Council Environment, March 11, 1999; accessed at http://www.consilium.europa.eu/uedocs/cms_data/docs/pressdata/en/envir/ACF87.htm on February 12, 2009.
20. 2194th Council meeting, Environment, Luxembourg, June 24, 1999; accessed at http://www.consilium.europa.eu/uedocs/cms_data/docs/pressdata/en/envir/ACF5B.htm#_Toc455373219 on February 12, 2009. Also see "Germany Vetoes End-of-Life Vehicles Directive," edie newsroom (Ireland), June 25, 1999, at http://www.edie.net/ireland/news/news_story.asp?id=1337&title=Germany+vetoes+End-of-Life+Vehicles+Directivechannel=0; accessed on February 10, 2007. Stephanie John and Daniela Schwarzer (2006: 18–19) attribute the German government's position to the impact on Chancellor Schroeder, a former VW Supervisory Board member while Minister President of Lower Saxony, of lobbying by Ferdinand Piëch, CEO of VW, in his role as President of the European Automobile Manufacturers' Association, ACEA. Wurzel (2000) also makes this point.
21. The trade-off involved German support for UK opposition to a proposal to give artists the right to royalties for their works sold throughout the single European market. Art dealers and auction houses in the United Kingdom opposed the measure for fear it would damage the art market in the United Kingdom. The

Council of Ministers and EP ultimately adopted a revised version of the Resale Right Directive (which made concessions to British objections) on September 27, 2001. See Philip Ward and Grahame Danby, "Artist's resale right," standard note SN/HA/4781, Home Affairs section, House of Commons Library; accessed at www.parliament.uk/briefing-papers/SN04781.pdf on August 12, 2009.

22. For the Council's July 1999 common position, see http://www.consilium.europa.eu/uedocs/cms_data/docs/pressdata/en/gena/10390.en9.htm#_Toc457986427. For the conflict within the Council caused by the German position and the Germany-UK trade-off, see Renée Cordes, "Finns Vow to Push for Deal on Car Recycling," *European Voice* (July 1, 1999); accessed at http://www.europeanvoice.com/article/imported/finns-vow-to-push-for-deal-on-car-recycling/38939.aspx on February 12, 2007. Ironically, Finland was later one of the governments against which the European Commission brought a case before the ECJ in 2003 for failing to fulfill its obligation to adopt provisions to implement the ELV rules by April 21, 2002, as required by the directive. See Case C-293/03, *Commission of the European Communities v. Republic of Finland*, OJC 213 of September 6, 2003, p. 18.
23. See "Agreement on Recycling End-of-Life Vehicles," European Information Service, *European Report* No. 2427, July 24, 1999; and "Germany Snubbed as Agreement Reached on Scrap Cars," *Automotive Environment Analyst* (September 1, 1999) (both accessed via LexisNexis Academic database).
24. Debates of the European Parliament, End-of-life vehicles second reading debate, February 3, 2000, Brussels; accessed at http://www.europarl.europa.eu/sides/getDoc.do?pubRef=-//EP//TEXT+CRE+20000203+ITEM-001+DOC+XML+V0//EN on February 12, 2007. Hereafter "EP, ELV debate."
25. EP, ELV debate.
26. Ibid.
27. Comments of John Bowis, PPE (British Conservative), ELV debate.
28. Comments of Roger Helmer, PPE (British Conservative), ELV debate.
29. Comments of Béatrice Patrie, PSE, EP, ELV debate.
30. Ibid.

31. Comments of Torben Lund, PSE, EP, ELV debate.
32. The overall vote (Amendment 38, 2nd part) was 180 for and 335 against, with 12 abstentions. Only 4 PSE members supported the amendment; 157 opposed. *Official Journal of the European Communities* (October 27, 2000), C 309/26–27.
33. For the text of the Lange amendment (Amendment 45) and the EP's other proposed second-reading amendments, see *Official Journal of the European Communities* (October 27, 2000), C 309/62–70, accessed at http://eur-lex.europa.eu/LexUriServ/LexUriServ.do?uri=OJ:C:2000:309:0062:0070:EN:PDF on February 12, 2007. For the outcome of the vote on the amendment, see *Official Journal* (October 27, 2000), C 309/28–29.
34. Tally by party and country based on author's calculations from the outcome of the vote on the Lange amendment provided in *Official Journal* (October 27, 2000), C 309/28–29.
35. This characterization refers to the committee structure of the fifth EP, 1999–2004.
36. This interpretation does not contradict the British government's acquiescence to the auto industry during the transition to the new EU directive, discussed in the section on "Implementation of the EU Directive and the Persistence of Industry Lobbying."
37. One Green MEP and a single Liberal voted in favor; 1 Green abstained. *Official Journal of the European Communities* (October 27, 2000), C 309/28–29.
38. Comments of Dirk Sterckx (ELDR, Flemish Liberal Democrat), ELV debate.
39. Comments of Alexander de Roo (Greens/ALE [European Free Alliance]), vice-chair of the Committee on the Environment, Public Health and Consumer Policy, ELV debate.
40. Comments of Hiltrud Breyer (Greens/ALE), ELV debate.
41. Comments of Marie Ann Isler Béguin (Greens/ALE), member of the Committee on the Environment, Public Health and Consumer Policy, ELV debate.
42. Comments of Jonas Sjöstedt (GUE [European United Left]/Nordic Green Left), member of the Committee on the Environment, Public Health and Consumer Policy, ELV debate.
43. Comments of Torben Lund (PES), member of the Committee on the Environment, Public Health and Consumer Policy, ELV debate.

44. Comments of European Environment Commissioner Margot Wallström, EP, second-reading, ELV debate, February 3, 2000.
45. See Anne Rasmussen, "The Role of the European Commission in Co-Decision—A Strategic Facilitator Operating in a Situation of Structural Disadvantage," *European Integration online Papers (EIoP)* Vol. 7, No. 10 (2003), http://eiop.or.at/eiop/texte/2003–010a.htm.
46. Commission of the European Communities, "Opinion of the Commission pursuant to Article 251 (2) (c) of the EC Treaty, on the European Parliament's amendments to the Council's Common Position Regarding the Proposal for a Directive of the European Parliament and of the Council on End of Life Vehicles," Brussels, March 16, 2000, COM (2000) 166 final, found at http://eur-lex.europa.eu/LexUriServ/LexUriServ.do?uri=COM:2000:0166:FIN:EN:PDF on August 17, 2009.
47. A European Parliament Policy Department study noted that "manufacturers have been extremely resistant to incurring extra costs prior to the date on which this was made mandatory by the Directive, and in such cases, final owners continued to have to bear any costs during the transition period." However, the study points out that some of the conflict was mitigated by a rise in the value of scrap metal during this period. See European Parliament (2007: vii).
48. These costs include removal and fire department response to the common practice of burning abandoned cars. See Paul Brown, "Illegal Dumping Fear Over Scrapped Cars: EU Rule Means Increased Cost for Getting Rid of Old Bangers," *The Guardian*, February 2, 2002. On the problem of abandoned vehicles in the United Kingdom, also see Daniel Attwood, "Motorists Will Not Pay for Disposal of Old Cars," *The Irish Times*, September 21, 2005 (Motors section, p. 1), which reports that "[i]n 2000, 23 per cent of ELVs—or 350,000 cars—were found abandoned in Britain"; both accessed via Lexis-Nexis Academic at http://www.lexisnexis.com.ezproxy.lib.ou.edu/hottopics/lnacademic/ on August 12, 2009. The Department of Trade and Industry estimated that there were 290,000 cars abandoned in 2001/2 and 310,000 in 2002/3. See European Parliament (2007: 53).
49. See "Cost of Scrapping Old Cars Will Fall to Motorists—Not Industry," *Bristol Evening Post*, June 15, 2002; and Matthew

George, "Paying the Price of Throwaway Society; Government Cave-in Will Drive Up Cost of Scrapping Old Vehicles," *Western Daily Press*, June 22, 2002; both accessed via Lexis-Nexis Academic at http://www.lexisnexis.com.ezproxy.lib.ou.edu/hottopics/lnacademic/ on August 12, 2009.

50. David Gow, "Industry Pleads for Delay Over eu Car Law," *The Guardian*, September 3, 2001; accessed via Lexis-Nexis Academic at http://www.lexisnexis.com.ezproxy.lib.ou.edu/hottopics/lnacademic/ on August 12, 2009.
51. See comments of Martin O'Neill (Labour, Ochil), House of Commons debate, End of Life Vehicles Directive, May 9, 2002; Hansard HC Deb 09 May 2002 vol 385 cc137-74WH; accessed on August 14, 2009, at http://hansard.millbanksystems.com/westminster_hall/2002/may/09/end-of-life-vehicles-directive.
52. Comments of Julie Kirkbride (Conservative, Bromsgrove), House of Commons debate, End of Life Vehicles Directive, May 9, 2002; Hansard HC Deb 09 May 2002 vol 385 cc137-74WH; accessed on August 14, 2009, at http://hansard.millbanksystems.com/westminster_hall/2002/may/09/end-of-life-vehicles-directive.
53. Mr. John Taylor (Conservative, Solihull), House of Commons debate, End of Life Vehicles Directive, May 9, 2002.
54. Mr. Philip Hammond (Conservative, Runnymeade and Weybridge), House of Commons debate, End of Life Vehicles Directive, May 9, 2002.
55. Brian Wilson, Ministry for Industry and Energy, House of Commons debate, End of Life Vehicles Directive, May 9, 2002.

## 5 Regulatory Trade-offs and Outcomes

1. Following successive treaty revisions, the principle is now articulated in Article 191(2) of Title XX ("Environment") of the 2009 Lisbon Treaty, which states that "Union policy on the environment…shall be based on the precautionary principle and on the principles that preventive action should be taken, that environmental damage should as a priority be rectified at source and that the polluter should pay."
2. President Reagan's Executive Order 12,291 required agencies to submit their regulatory impact analyses to the Office of

Management and Budget for review and stipulated that regulatory action should not be undertaken if the societal benefits did not "outweigh" the costs. President Clinton's Executive Order 12,866 revised this strict cost-benefit criterion, establishing that the benefits of regulation "justify" the costs. In addition, the Clinton administration expanded the range of rules (which involve the use of regulatory agency authority to implement laws passed by Congress) requiring economic impact analysis subject to OMB review. See the EPA's website for "Statutory and Administrative Requirements for Economic Analysis of Regulations" at http://www.epa.gov/ttn/ecas/econdata/Rmanual2/2.2.html.

3. The US regulation is the Regulatory Flexibility Act of 1980. The text is available on the US Small Business Association Office of Advocacy website at http://archive.sba.gov/advo/laws/regflex.html. European Commission impact assessment guidelines call for application of an "SME Test" to consider "whether SMEs are disproportionately affected or disadvantaged compared to large companies and if so, options should cover alternative mechanisms and flexibilities in approach that might help SMEs to comply." See the European Commission Enterprise DG website at http://ec.europa.eu/enterprise/policies/sme/small-business-act/sme-test/index_en.htm.
4. See "Inter-Institutional Common Approach to Impact Assessment (IA)," November 2005, at http://ec.europa.eu/governance/better_regulation/documents/ii_common_approach_to_ia_en.pdf.
5. The effectiveness of regulatory impact assessment is a subject of debate for both the United States and the EU cases. For an account that emphasizes poor compliance with the Clinton administration's executive order, see Robert W. Hahn et al., "Assessing Regulatory Impact Analyses: The Failure of Agencies to Comply with Executive Order 12,866," *Harvard Journal of Law & Public Policy* Vol. 23, No. 3 (1999–2000): 859–885. For a critical account of the quality of the European Commission's impact assessments, see Renda (2006), especially pp. 61–70.
6. Among others, Khanna (2001: 292) points out that "cut backs in the regulatory budget of the U.S. Environmental Protection Agency (USEPA) have limited its ability to monitor and enforce policies since the mid 1980's."

7. Segerson and Miceli (1998) argue that regulators have incentives to pursue voluntary agreements even with lower abatement levels due to the cost of mandatory regulation.
8. Khanna (2001: 318) suggests that a review of studies of voluntary approaches to environmental protection "demonstrates the potential for firms to be able to preempt regulation with a very modest amount of voluntary abatement that might be less than that would have been imposed by the regulator."
9. As Alberini and Segerson (2002: 170) argue, "When the background threat" (of mandatory regulation) "is weak…then the level of abatement that emerges from the agreement is likely to be much lower." These authors find generally that "the existence of a strong regulatory threat" (178) is an essential feature of effective voluntary regulation, since it increases the regulator's bargaining power. Khanna (2001: 318) concurs, as do Segerson and Miceli (1998: 128).
10. Actions were dropped as governments transposed the directive, with infringement proceedings against five (Belgium, France, Italy, Portugal, and the United Kingdom). *European Report*, November 8, 2007; accessed via LexisNexis Academic database.
11. As described by the EPA, " Hydraulic fracturing is a process in which large volumes of water, sand and chemicals are injected at high pressures to extract oil and natural gas from underground rock formations. The process creates fractures in formations such as shale rock, allowing natural gas or oil to escape into the well and be recovered." See the EPA news release announcing its hydraulic fracturing study plan at http://yosemite.epa.gov/opa/admpress.nsf/d0cf6618525a9efb85257359003fb69d/26195e235a35cb3885257831005fd9cd!OpenDocument; accessed on July 30, 2011.
12. The agreement may be found at http://www.epa.gov/ogwdw000/uic/pdfs/moa_uic_hyd-fract.pdf; accessed on July 30, 2011. Diesel contains toxic constituents, including benzene, a known carcinogen.
13. While eight of these firms complied with the request, Halliburton first declined to provide information, but then responded to a subpoena from the EPA. See "Hydraulic Fracturing Information Request" at http://water.epa.gov/type/groundwater/uic/class2/hydraulicfracturing/index.cfm; accessed on July 30, 2011.

14. See Mark Clayton, "EPA to Natural Gas Companies: Give Details on 'Fracking' Chemicals," *The Christian Science Monitor*, September 9, 2010; accessed on July 30, 2011, via Lexis Nexis Academic database at http://www.lexisnexis.com.ezproxy.lib.ou.edu/hottopics/lnacademic/
15. See "US House Probe Accuses 12 Firms of Illegally Fracturing NatGas Wells," *Natural Gas Week*, February 7, 2011; accessed via Lexis Nexis Academic database at http://www.lexisnexis.com.ezproxy.lib.ou.edu/hottopics/lnacademic/ on July 30, 2011.
16. See Nick Snow, "Congressmen to Reintroduce Bill to Federally Regulate Fracing," *Oil & Gas Journal*, February 28, 2011; accessed via Lexis Nexis Academic database at http://www.lexisnexis.com.ezproxy.lib.ou.edu/hottopics/lnacademic/ on July 30, 2011. H.R. 1084, the Fracturing Responsibility and Awareness of Chemicals Act of 2011, was referred to the Subcommittee on Environment and the Economy of the House Committee on Energy and Commerce on March 21, 2011. See http://www.govtrack.us/congress/bill.xpd?bill=h112–1084.
17. See Dave Michaels, "Some Drillers Say They'll Disclose Fracking Chemicals," *Dallas Morning News*, April 13, 2011; accessed via Lexis Nexis Academic database at http://www.lexisnexis.com.ezproxy.lib.ou.edu/hottopics/lnacademic/ on July 30, 2011. Several states, including California and Texas, also passed or were considering hydraulic fracturing chemicals disclosure laws in 2011.
18. See "A Non-Paper with commentary on the suggestions received by the High Level Group of Independent Stakeholders on Administrative Burden for the hearing on 12 April 2010 on unnecessary administrative burden for environmental legislation"; accessed at http://ec.europa.eu/enterprise/policies/smart-regulation/administrative-burdens/high-level-group/files/abr_hlg_120410_non-paper_en.pdf on August 19, 2010.
19. See "Non-Paper," Commission commentary.
20. "Proposal for a Regulation of the European Parliament and the Council concerning the placing on the market and use of biocidal products," COM/2009/0267 final; accessed at http://eur-lex.europa.eu/LexUriServ/LexUriServ.do?uri=CELEX:DKEY=496597:EN:NOT on August 19, 2010.

21. Biocidal products regulation, Explanatory Memorandum, section 1.2. The Commission also noted (section 2) that "the data requirements are seen as particularly strict and in some cases as non-proportionate or inflexible."
22. See, inter alia, the joint letter of major European environmental groups to the Belgian presidency of the Environment Council coordinated by PAN Germany Pesticides Action Network; available at http://www.pan-germany.org/download/biocides/Biocides_Jointletter_Council_20101214.pdf.
23. The reference here is to exclusion criterion (c) in Article 5, part 1 of the proposed regulation, according to which substances that are carcinogenic, mutagenic, toxic to the reproductive system, or disruptive to the endocrine system could nonetheless be authorized if "it is shown that not including the active substance...would cause disproportionate negative impacts when compared with the risk to human health or the environment arising from the use of the substance and that there are no suitable alternative substances or technologies."
24. See Press Release, 3061st Council Meeting, Environment, Brussels, December 20, 2010, p. 8; accessed on July 31, 2011, at http://consilium.europa.eu/uedocs/cms_data/docs/pressdata/en/envir/118652.pdf.

# Bibliography

Alberini, Anna and Kathleen Segerson. "Assessing Voluntary Programs to Improve Environmental Quality," *Environmental and Resource Economics* 22, No. 1–2 (June 2002): 157–184.

Andrews, Richard N. L. "Environmental Regulation and Business 'Self-Regulation'," *Policy Sciences* 31 (1998): 177–197.

Asp, Nils-Georg and Susanne Bryngelsson. "Health Claims in the Labelling and Marketing of Food Products: The Swedish Food Sector's Code of Practice in a European Perspective," *Scandinavian Journal of Food and Nutrition* 51, No. 3 (2007): 107–126.

Atlas, Mark. "TRI to Communicate: Public Knowledge of the Toxics Release Inventory," *Social Science Quarterly* 88, No. 2 (2007): 555–572.

Bailer, Stefanie. "A Political Economy of Preferences in EU Negotiations: An Analysis of the Determinants of Positions in the Council of Ministers," Center for Comparative and International Studies, University of Zurich (2008). Accessed at http://www.cis.ethz.ch/events/past_events/PEIO2008/Bailer_EU.Council.Preferences on August 16, 2010.

Bergeson, Lynn L., Lisa M. Campbell, and Lisa Rothenberg. "TSCA and the Future of Chemical Regulation," *EPA Administrative Law Reporter* 15, No. 4 (April 2000): 1–23.

Bergner, Danielle M. "The Electronic Waste Recycling Act of 2003: California's Response to the Electronic Waste Crisis," *Marquette Law Review* 88 (2004): 377–389.

BEUC. "BEUC Comments on the Request for Information in View of the Revision of Council Directive 90/496/EEC on Nutrition Labelling" (March 7, 2003). Accessed at http://ec.europa.eu/food/food/labellingnutrition/nutritionlabel/beuc_en.pdf

Beyers, Jan. "Gaining and Seeking Access: The European Adaptation of Domestic Interest Associations," *European Journal of Political Research* 41, No. 5 (August 2002): 585–612.

Blim, John M. "Free Speech and Health Claims Under the Nutrition Labeling and Education Act of 1990: Applying a Rehabilitated *Central Hudson* Test for Commercial Speech," *Northwestern University Law Review* 88, No. 2 (1994): 733–768.

Bone, Paula Fitzgerald and Karen Russo France. "Qualified Health Claims on Package Labels," *Journal of Public Policy & Marketing* 28, No. 2 (Fall 2009): 253–258.

Bouwen, Pieter. "Corporate Lobbying in the European Union: The Logic of Access," *Journal of European Public Policy* 9, No. 3 (2002): 365–390.

———. "The Logic of Access to the European Parliament: Business Lobbying in the Committee on Economic and Monetary Affairs," *Journal of Common Market Studies* 42, No. 3 (2004): 473–495.

Bradley, Michele M. "The States' Role in Regulating Food Labeling and Advertising: The Effect of the Nutrition Labeling and Education Act of 1990," *Food & Drug Law Journal* 49 (1994): 649–674.

Brickman, Ronald, Sheila Jasanoff, and Thomas Ilgen. *Controlling Chemicals: The Politics of Regulation in Europe and the United States* (Ithaca and London: Cornell University Press, 1985).

Bulterman, Mielle. "Selected Legislation and Jurisprudence, Court of Justice of the European Communities, CJEC 2003/1, Case C-221/00, Commission/Austria," *European Journal of Health Law* 10 (2003): 65–66.

Carver, Krista Hessler. "A Global View of First Amendment Constraints on FDA," *Food & Drug Law Journal* 63 (2008): 151–215.

Cecot, Caroline, Robert Hahn, Andrea Redna, and Lorna Schrefler. "An Evaluation of the Equality of Impact Assessment in the European Union with Lessons for the US and the EU," *Regulation & Governance* 2 (2008): 405–424.

ChemSec (International Chemical Secretariat). "Cry Wolf: Predicted Costs by Industry in the Face of New Regulations," Report 6:04 (April 2004). Available at: http://www.chemsec.org/images/stories/news_publications/Cry_wolf_report_040422.pdf

Childs, Nancy M. "Public Policy Approaches to Establishing Health Claims for Food Labels: An International Comparison," *British Food Journal* 100, No. 4 (1998): 191–200.

Christoforou, Theofanis. "The Precautionary Principle, Risk Assessment, and the Comparative Role of Science in the European Community and the US Legal Systems," in Norman J. Fig and Michael J. Faure, eds., *Green Giants? Environmental Policies of the United States and the European Union* (Cambridge and London: MIT Press, 2004), pp. 17–51.

Cichowski, Rachel A. "Integrating the Environment: The European Court and the Construction of Supranational Policy," *Journal of European Public Policy* 5, No. 3 (September 1998): 387–405.

Close, Cavan O'Connor and Dominic J. Mancini. "Comparison of U.S. and European Commission Guidelines on Regulatory Impact Assessment/Analysis," Industrial Policy and Economic Reforms Papers No. 3, Enterprise and Industry Directorate-General, European Commission (April 2007). Accessed at http://ec.europa.eu/enterprise/newsroom/cf/_getdocument.cfm?doc_id=4188 on August 19, 2010.

Coen, David. "The Evolution of the Large Firm as a Political Actor in the European Union," *Journal of European Public Policy* 4, No. 1 (1997): 91–108.

———. "Environmental and Business Lobbying Alliances in Europe: Learning from Washington?" in D. Levy and P. Newell, eds., *Business in International Environmental Governance: A Political Economy Approach* (Cambridge, MA: MIT Press, 2004), pp. 197–220.

———. "Empirical and Theoretical Studies in EU lobbying," *Journal of European Public Policy* 14, No. 3 (April 2007): 333–345.

Collins, Ken, Charlotte Burns, and Alex Warleigh. "Policy Entrepreneurs: The Role of European Parliament Committees in the Making of EU Policy," *Statute Law Review* 19, No. 1 (1998): 1–11.

Commission of the European Communities. "Communication from the Commission on the Review of the Community Strategy for Waste Management and Draft Council Resolution on Waste Policy," COM (96) 399 final, 30.07.1996 (1996). Available at Archive of European Integration, University of Pittsburgh, http://aei.pitt.edu/4116/.

———. "Proposal for a Council Directive on End of Life Vehicles," COM (97) 358 final, July 9, 1997.

———. "White Paper on Food Safety," Brussels, COM (1999) 719 final, January 12, 2000. Accessed at http://ec.europa.eu/dgs/health_consumer/library/pub/pub06_en.pdf on August 24, 2010.

———. "Discussion Paper on Nutrition Claims and Functional Claims," Prepared by Directorate General Health and Consumer Protection,

SANCO/1341/2001 (2001). Accessed at http://ec.europa.eu/food/food/labellingnutrition/claims/background_en.htm on August 24, 2010.

———. "Proposal for a Regulation of the European Parliament and of the Council on Nutrition and Health Claims Made on Foods," Brussels, 16.7.2003 COM(2003) 424 final, July 16, 2003.

———. "A Strategic Review of Better Regulation in the European Union," Communication from the Commission to the European Parliament, the Council, the European Economic and Social Committee and the Committee of the Regions, Brussels, COM(2006) 689 final, November 14, 2006.

———. "Report from the Commission to the Council, the European Parliament, the European Economic and Social Committee and the Committee of the Regions on the Implementation of Directive 2000/53/EC on End-of-Life Vehicles for the period 2002–2005," Brussels, COM(2007) 618 final, October 17, 2007.

———. "Second Strategic Review of Better Regulation in the European Union," Communication from the Commission to the European Parliament, the Council, the European Economic and Social Committee and the Committee of the Regions, Brussels, COM(2008) 32 final, January 30, 2008.

Corporate Europe Observatory. "Bulldozing REACH—the Industry Offensive to Crush EU Chemicals Regulation." Retrieved from http://archive.corporateeurope.org/docs/lobbycracy/BulldozingREACH.pdf. March 2005.

Council of the European Union. "Common Position Adopted by the Council with a View to the Adoption of a Regulation of the European Parliament and of the Council on Nutrition and Health Claims Made on Foods," Brussels, (9858/05), November 15, 2005.

Coutrelis, Nicole. "European Union Food Law Update," *Journal of Food Law & Policy* (2005): 219.

Davies, J. Clarence and Jan Mazurek. *Pollution Control in the United States: Evaluating the System*. Resources for the Future Press, 1998.

Dunlop, Claire. "GMOs and Regulatory Styles," *Environmental Politics* 9, No. 2 (Summer 2000): 149–155.

Echols, Marsha A. "Food Safety Regulation in the European Union and the United States: Different Cultures, Different Laws," *Columbia Journal of European Law* 4 (1998): 525–543.

Eising, Rainer. "Institutional Context, Organizational Resources and Strategic Choices: Explaining Interest Group Access in the European Union," *European Union Politics* 8, No. 3 (2007): 329–362.

Environmental Defense Fund. "Toxic Ignorance: The Continuing Absence of Basic Health Testing for Top-Selling Chemicals in the United States," New York: Environmental Defense Fund (1997).

Environmental Working Group, Chemical Industry Archives. "Responsible? Care?: As Bad News Mounts and Polls Head South, Chemical Companies Spend Millions on 'Public Perception'," Available at: http://www.chemicalindustryarchives.org/dirtysecrets/responsiblecare/1.asp. Updated March 2009. (2009a).

———. "Voluntary Chemical Safety Testing," Accessed at http://www.chemicalindustryarchives.org/dirtysecrets/testing/1.asp on September 2009. (2009b).

European Food Safety Authority. "EFSA Conference on Nutrition and Health Claims," EFSA Meeting Summary Report, November 8–10, 2006, Bologna, Italy (EFSA, February 2008). Available at http://www.medinfo.hu/new3/konyvtar//bookshop/EFSA%20Conference%20on%20Nutrition%20and%20Health%20Claims.pdf

European Parliament (EP). "Report on the Proposal for a Council Directive on End of Life Vehicles," COM(97)0358–C4-0639 / 97–97/0194(SYN)), Committee on the Environment, Public Health and Consumer Protection, 1999.

———. "Report on the Joint Text Approved by the Conciliation Committee for a European Parliament and Council Directive on End-of-Life Vehicles (C5-0258/2000-1997/0194(COD))," European Parliament Delegation to the Conciliation Committee, Final A5-0212/2000, July 15, 2000.

———. "Report on the Commission White Paper on Strategy for a Future Chemicals Policy (COM(2001) 88-C5-0258/2001- 2001/2118(COS))," Committee on the Environment, Public Health and Consumer Policy, Final A5-0356/2001, October 17, 2001.

———. "Report on the Proposal for a Regulation of the European Parliament and of the Council on Nutrition and Health Claims Made on Foods," COM(2003 0424–C5-0329/2003– 2003/0165(COD)), Commission on the Environment, Public Health and Food Safety, Adriana Poli Bortone, Rapporteur. Final A6-0128/2005, 2005.

European Parliament (EP). "End of Life Vehicles (ELV) Directive: An Assessment of the Current State of Implementation by Member States," Policy Department, Economic and Scientific Policy, IP/A/ENVI/ST/2006-29 (March 2007). Available at: http://www.europarl.europa.eu/comparl/envi/pdf/externalexpertise/end_of_life_vehicles.pdf.

Fairbrass, J. and Jordan, A. "Protecting Biodiversity in the European Union: National Barriers and European Opportunities?" *Journal of European Public Policy* 8, No. 4 (2001): 499–518.

Falkner, Robert. "The Political Economy of 'Normative Power' Europe: EU Environmental Leadership in International Biotechnology Regulation," *Journal of European Public Policy* 14, No. 4 (June 2007): 507–526.

Fealk-Stickler, Lisa M. "Regulating the Regulators: The Impact of FDA Regulation on Corporations' First Amendment Rights," *John Marshall Law Review* 39 (2005–2006): 95–114.

Gaynor, Kevin. "The Toxic Substances Control Act: A Regulatory Morass," *Vanderbilt Law Review* 30 (1977): 1149–1195.

General Accounting Office. "Toxic Substances Control Act: Preliminary Observations on Legislative Changes to Make TSCA More Effective," Testimony before the Subcommittee on Toxic Substances, Research and Development, Committee on Environment and Public Works, U.S. Senate, Statement of Peter F. Guerrero, Director, Environmental Protection Issues, July 13, 1994.

Givel, Michael. "Motivation of Chemical Industry Social Responsibility through Responsible Care," *Health Policy* 81, No. 1 (April 2007): 85–92.

Glicksman, Robert and Christopher H. Schroeder. "EPA and the Courts: Twenty Years of Law and Politics," *Law & Contemporary Problems* 54 (1991): 249–309.

Golub, Jonathan. "State Power and Institutional Influence in European Integration: Lessons from the Packaging Waste Directive," *Journal of Common Market Studies* 34, No. 3 (September 1996): 313–339.

———. *Global Competition and EU Environmental Policy* (London: Routledge, 1998).

Government Accountability Office (GAO). "Chemical Regulation: Options Exist to Improve EPA's Ability to Assess Health Risks and Manage its Chemical Review Program," Report to Congressional Requesters, GAO-05-458, June 2005.

———. "Chemical Regulation: Options for Enhancing the Effectiveness of the Toxic Substances Control Act," Testimony before the Subcommittee on Commerce, Trade, and Consumer Protection, Committee on Energy and Commerce, House of Representatives, February 26, 2009. (2009a).

———. "Chemical Regulation: Observations on Improving the Toxic Substances Control Act," Testimony before the Committee on Environment and Public Works, U.S. Senate, December 2, 2009. (2009b).

General Accounting Office. "Toxic Substances Control Act: EPA's Limited Progress in Regulating Toxic Chemicals," Testimony before the Subcommittee on Toxic Substances, Research and Development, Committee on Environment and Public Works, U.S. Senate, Statement of Peter F. Guerrero, Director, Environmental Protection Issues, Resources, Community, and Economic Development Division, May 17, 1994. Accessed at http://archive.gao.gov/t2pbat3/151661.pdf

Guth, Joseph H., Richard A. Denison, and Jennifer Sass. "Background Paper for Reform No. 5 of the Louisville Charter: Require Comprehensive Safety Data for All Chemicals." Accessed at http://www.louisvillecharter.org/paper.safetydata.shtml on July 14, 2005.

Haemer, Robert B. "Reform of the Toxic Substances Control Act: Achieving Balance in the Regulation of Toxic Substances," *The Environmental Lawyer* 6 (1999–2000): 99–134.

Hanan, Andrew. "Pushing the Environmental Regulatory Focus a Step Back: Controlling the Introduction of New Chemicals Under the Toxic Substances Control Act," *American Journal of Law & Medicine* 18, No. 4 (1992): 395–421.

Hawkes, Corrina. *Nutrition Labels and Health Claims: The Global Regulatory Environment* (Geneva: World Health Organization, 2004).

Hemphill, Thomas A. "Corporate Environmentalism and Self-Regulation: Keeping Enforcement Agencies at Bay," *Journal of Environmental Regulation* 3, No. 2 (1993): 145–154.

Hojnacki, Marie, and David C. Kimball. "Organized Interests and the Decision of Whom to Lobby in Congress," *American Political Science Review* 92 (1998): 775–790.

Hunter, Beatrice Trum. "Relaxing the Rules on Food Labels," *Consumers' Research Magazine* 79, No. 4 (April 1996): 18–21.

John, Stephanie and Daniela Schwarzer. "Industry Lobbying within the European Union: Actors, Strategies, and Trends in the Multi-Level

System," American Institute for Contemporary German Studies, Policy Report 24, 2006.

Kamieniecki, Sheldon. *Corporate America and Environmental Policy: How Often Does Business Get its Way?* (Stanford, CA: Stanford University Press, 2006).

Kapstein, Ethan. "Resolving the Regulator's Dilemma: International Coordination of Banking Regulations," *International Organization* 43, No. 2 (Spring 1989): 323–347.

Katan, Martijn B. and Nicole M. de Roos. "Toward Evidence-Based Health Claims for Foods," *Science* 299, No. 5604 (January 10, 2003): 206–207.

Keading, Michael. "Rapporteurship Allocation in the European Parliament: Information or Distribution," *European Union Politics* 5, No. 3 (2004): 353–371.

Keane, Alice D. "Federal Regulation of Solid Waste Reduction and Recycling," *Harvard Journal on Legislation* 29 (1992): 251–279.

Kelemen, R. Daniel. "Regulatory Federalism: EU Environmental Regulation in Comparative Perspective," *Journal of Public Policy* 20, No. 2 (May–August 2000): 133–167.

Khanna, Madhu. "Non-Mandatory Approaches to Environmental Protection," *Journal of Economic Surveys* 15, No. 3 (July 2001): 291–324.

Kibert, Nicole C. "Extended Producer Responsibility: A Tool for Achieving Sustainable Development," *Journal of Land Use* 19, No. 2 (Spring 2004): 503–523.

King, Andrew A. and Michael J. Lenox. "Industry Self-Regulation without Sanctions: The Chemical Industry's Responsible Care Program," *Academy of Management Journal* 43, No. 4 (August 2000): 698–716.

Kohler-Koch, Beate. "Organized Interests in the EC and the European Parliament," *European Integration Online Papers* 1: 9 (1997).

Konz, Raymond J. "The End-of-Life Vehicle (ELV) Directive: The Road to Responsible Disposal," *Minnesota Journal of International Law* 18 (2009): 431–457.

Kossdorff, Katharina. "The New Austrian Food Safety and Consumer Protection Act," *Ernährung/Nutrition* 30, No. 12 (2006): 525–532.

Krämer, Ludwig. "The Roots of Divergence: A European Perspective," in Norman J. Vig and Michael G. Faure, eds., *Green Giants? Environmental Policies of the United States and the European Union* (Cambridge and London: MIT Press, 2004), pp. 53–72.

Krasny, Leslie T. "Three Options: FDA Outlines Alternative Approaches for Regulating Qualified Health Claims," *Food Processing* 65, No. 2 (February 1, 2004).

Kurzer, Paulette. "Transatlantic Risk Perceptions, Public Health, and Environmental Concerns: Coming Together or Drifting Apart?" in Nicolas Jabko and Craig Parsons, eds., *The State of the European Union, Volume 7: With US or Against US? European Trends in American Perspective* (New York: Oxford University Press, 2005), pp. 299–324.

Löfstedt, Ragnar E. "The Swing of the Regulatory Pendulum in Europe: From Precautionary Principle to (Regulatory) Impact Analysis," *The Journal of Risk and Uncertainty* 28, No. 3 (2004): 237–260.

Löfstedt, Ragnar E. and David Vogel. "The Changing Character of Regulation: A Comparison of Europe and the United States," *Risk Analysis* 21, No. 3 (2001): 399–405.

Lowell Center for Sustainable Production. "The Promise and Limits of the United States Toxic Substances Control Act," Lowell, MA, October 10, 2003.

Lucas, Rainer. "End-of-Life Vehicle Regulation in Germany and Europe—Problems and Perspectives," Wuppertal Papers, No. 113 (March 2001), Wuppertal Institute for Climate, Environment and Energy, Wuppertal, Germany. Available at: http://www.wupperinst.org.

Lyon, Thomas P. and John W. Maxwell. "'Voluntary' Approaches to Environmental Regulation: A Survey," (January 1999). Available at: SSRN: http://ssrn.com/abstract=147888 or DOI: 10.2139/ssrn.147888

Mahoney, C. "The Power of Institutions: State and Interest-Group Activity in the European Union," *European Union Politics* 5, No. 4 (2004): 441–466.

Marks, G. and McAdam, D. "Social Movements and the Changing Structure of Political Opportunity in the European Union," *West European Politics* 19, No. 2 (1996): 249–278.

Maxwell, John W., Thomas P. Lyon, and Steven C. Hackett. "Self-Regulation and Social Welfare: The Political Economy of Corporate Environmentalism," *Journal of Law & Economics* 43 (October 2000): 583–617.

Michaels, Mara A. "FDA Regulation of Health Claims Under the Nutrition Labeling and Education Act of 1990: A Proposal for a Less Restrictive Scientific Standard," *Emory Law Journal* 44 (1995): 319–357.

Nicol, Scott and Shirley Thompson. "Policy Options to Reduce Consumer Waste to Zero: Comparing Product Stewardship and Extended Producer Responsibility for Refrigerator Waste," *Waste Management & Research* 25, No. 3 (2007): 227–233.

Noah, Lars and Barbara A. Noah. "Liberating Commercial Speech: Product Labeling Controls and the First Amendment," *Florida Law Review* 47 (1995): 63–112.

Ogushi, Yasuhiko and Milind Kandlikar. "Assessing Extended Producer Responsibility Laws in Japan," *Environmental Science & Technology Online* 41, No. 13 (July 1, 2007): 4502–4508.

O'Toole, Laurence J., Jr., Chilik Yu, James Cooley, Gail Cowie, Susan Crow, Terry DeMeo, and Stephanie Herbert. "Reducing Toxic Chemical Releases and Transfers: Explaining Outcomes for a Voluntary Program," *Policy Studies Journal* 25, No. 1 (March 1997): 11–26.

Oye, K. A. "The Precautionary Principle and International Conflict over Domestic Regulation: Mitigating Uncertainty and Improving Adaptive Capacity," *Water Science and Technology* 52, No. 6 (2005): 59–64.

Palmer, Elizabeth A. "'Food Fight' Ends, New Labels Should Be Easier to Digest," *Congressional Quarterly Weekly Report* 50, No. 48 (December 5, 1992): 3751.

Palmer, Karen, Wallace E. Oates, and Paul R. Portney. "Tightening Environmental Standards: The Benefit-Cost or the No-Cost Paradigm?" *Journal of Economic Perspectives* 9, No. 4 (Autumn 1995): 119–132.

Pappas, Clement Dimitri. "Maintaining a Level Playing Field: The Need for a Uniform Standard to Evaluate Health Claims for Foods and Dietary Supplements," *Food and Drug Law Journal* 57 (2002): 25–41.

Percival, Robert V. "Environmental Legislation and the Problem of Collective Action," *Duke Environmental Law & Policy Forum* 9 (Fall 1998): 9–27.

Pollack, M. A. "Representing Diffuse Interests in EC Policy-making," *Journal of European Public Policy* 4, No. 4 (1997): 572–590.

Porter, Michael and Class van der Linde. "Toward a New Conception of the Environment-Competitiveness Relationship," *Journal of Economic Perspectives* 9, No. 4 (Autumn 1995): 97–118.

Radaelli, Claudio M. "Whither Better Regulation for the Lisbon Agenda?" *Journal of European Public Policy* 14, No. 2 (March 2007): 190–207.

Rees, Joseph. "Development of Communitarian Regulation in the Chemical Industry," *Law & Policy* 19, No. 4 (October 1997): 477–528.

Renda, Andrea. *Impact Assessment in the EU: The State of the Art and the Art of the State*. Brussels: Centre for European Policy Studies, 2006.

Reynolds, Joel. "The Toxic Substances Control Act of 1976: An Introductory Background and Analysis," *Columbia Journal of Environmental Law* 4 (1977): 35–96.

Rogers, William. "The Lesson of the Owl and the Crows: The Role of Deception in the Evolution of the Environmental Statutes," *Journal of Land Use & Environmental Law* 4, No. 2 (Fall 1988): 377–388.

Rosendal, G. Kristin. "Governing GMOs in the EU: A Deviant Case of Environmental Policy-Making?" *Global Environmental Politics* 5, No. 1 (2005): 82–104.

Ruffell, Melanie. "UK Claims Watchdog Starts to Bark: Joint Health Claims Initiative (JHCI) Up and Running," *Nutrition Bulletin* 26: 59–60 (2001).

———. "Health Claims on Food—JHCI Update Substantiation Requirements," *Nutrition Bulletin* 28: 193–196 (2003).

Ruggerio, Cynthia. "Referral of Toxic Chemical Regulation Under the Toxic Substances Control Act: EPA's Administrative Dumping Ground," *Boston College Environmental Affairs Law Review* 17 (1989–1990): 75–122.

Sachs, Noah. "Planning the Funeral at the Birth: Extended Producer Responsibility in the European Union and the United States," *Harvard International Law Review* 30 (2006): 51–98.

Salzman, James. "Sustainable Consumption and the Law," *Environmental Law* 27 (1997): 1243–1294.

Samp, Richard A. "Courts Are Arriving at a Consensus on Food and Drug Administration Speech Regulation," *Food and Drug Law Journal* 58 (2003): 313–330.

Segerson, Kathleen and Thomas J. Miceli. "Voluntary Environmental Agreements: Good or Bad News for Environmental Protection?" *Journal of Environmental Economics and Management* 36 (1998): 109–130.

Selin, Henrik and Stacy D. VanDeveer. "Raising Global Standards: Hazardous Substances and E-Waste Management in the European Union," *Environment* 48, No. 10 (December 2006): 6–17.

Sharma, Lisa L., Stephen P. Teret, and Kelly D. Brownell. "The Food Industry and Self-Regulation: Standards to Promote Success and to

Avoid Public Health Failures," *American Journal of Public Health* 100, No. 2 (February 2010): 240–246.

Shörling, Inger. *REACH—The Only Planet Guide to the Secrets of Chemicals Policy in the EU. What Happened and Why?* The Greens/European Free Alliance in the European Parliament, April 2004.

Short, Megan. "Taking Back the Trash: Comparing European Extended Producer Responsibility and Take-Back Liability to U.S. Environmental Policy and Attitudes," *Vanderbilt Journal of Transnational Law* 37 (2004): 1217–1254.

Silverglade, Bruce and Ilene Ringel Heller. "Food Labeling Chaos: The Case for Reform," Center for Science in the Public Interest (2010). Accessed at http://www.cspinet.org/new/pdf/food_labeling_chaos_report.pdf on May 5, 2011.

Simmons, Peter and Brian Wynne. "Responsible Care: Trust, Credibility and Environmental Management," in Kurt Fischer and Johan Schot, eds., *Environmental Strategies for Industry* (Washington, DC: Island Press, 1993), pp. 201–226.

Sissell, Sara. "Responsible Care at 20," *Chemical Week* 170, No. 21 (July 7–14, 2008): 23–26.

Slesin, Louis. "Regulating Chemicals Under TOSCA," *Environment* 20, No. 2 (March 1978): 19–20.

Smith, Mitchell P. "All Access Points Are Not Created Equal: Explaining the Fate of Diffuse Interests in the EU," *British Journal of Politics and International Relations* 10, No. 1 (February 2008): 64–83.

———. "Single Market, Global Competition: Regulating the European Market in a Global Economy," *Journal of European Public Policy* 17, No. 7 (October 2010): 936–953.

Staudinger, Jeff and Gregory A. Keoleian. "Management of End-of-Life-Vehicles (ELVs) in the U.S.," Center for Sustainable Systems, University of Michigan, Report No. CSS01-01 (March 2001). Accessed at http://css.snre.umich.edu/css_doc/CSS01-01.pdf on August 19, 2010.

Steinborn, Steven. "The Times They Are a-Changing: Qualified Health Claims Are Opening New Doors as Marketers Strive to Communicate Their Product's Health Benefits to Consumers," *Prepared Foods* 173, No. 8 (August 1, 2004).

Steinborn, Steven B. and Kyra Todd. "The End of Paternalism: A New Approach to Food Labeling," *Food and Drug Law Journal* 54 (1999): 401–422.

Steinzor, Rena I. "Reinventing Environmental Regulation: The Dangerous Journey from Command to Self-Control," *Harvard Environmental Law Review* 22 (1998): 103–202.

Strasser, Kurt A. "Do Voluntary Corporate Efforts Improve Environmental Performance? The Empirical Literature," *Boston College Environmental Affairs Law Review* 35 (2008): 533–555.

Tanguay, Georges A., Paul Lanoie, and Jérôme Moreau. "Environmental Policy, Public Interest and Political Market," *Public Choice* 120 (2004): 1–27.

Toffel, Michael W. "The Regulatory and Judicial Roots of Product Stewardship in the United States," Haas School of Business Working Paper (September 27, 2002). Accessed at http://faculty.haas.berkeley.edu/toffel/papers/Regulatory%20&%20judicial%20roots%20of%20PS%20in%20US.pdf on February 9, 2007.

———. "The Growing Strategic Importance of End-of-life Product Management," *California Management Review* 45, No. 3 (Spring 2003): 102–129.

Togawa, Kenichi. "Background of the Automobile Recycling Law Enactment in Japan," *Environmental Economics and Policy Studies* 6 (2005): 271–283.

Tufet-Opi, Enrique. "Life After End of Life: The Replacement of End of Life Product Legislation by an European Integrated Product Policy in the EC," *Journal of Environmental Law* 14, No. 1 (2002): 33–60.

U.S. Congress, H.B. 4308 [109th]. "Child, Worker, and Consumer-Safe Chemicals Act of 2005." Accessed at http://www.govtrack.us/congress/billtext.xpd?bill=h109-4308.

U.S. Department of Health and Human Services, Food and Drug Administration. "Comments of American Association of Family and Consumer Sciences, American Cancer Society, American College of Preventive Medicine, Association of American Medical Colleges, Center for Science in the Public Interest, Consumer Federation of America, National Consumers League," Advanced Notice of Proposed Rulemaking, Food Labeling: Health Claims; Dietary Guidance, February 25, 2004.

Vedder, Hans. *Competition Law and Environmental Protection in Europe: Towards Sustainability?* (Netherlands: Europa Law Publishing, 2003).

Vladeck, David C. "Devaluing Truth: Unverified Health Claims in the Aftermath of *Pearson v. Shalala*," *Food & Drug Law Journal* 54: 535–553 (1999).

Wiener, Jonathan B. "Convergence, Divergence, and Complexity in US and European Risk Regulation," in Norman J. Vig and Michael G. Faure, eds., *Green Giants? Environmental Policies of the United States and the European Union* (Cambridge and London: MIT Press, 2004), pp. 73–109.

Wiener, Jonathan B. and Michael D. Rogers. "Comparing Precaution in the United States and Europe," *Journal of Risk Research* 5, No. 4 (2002): 317–349.

Winter, Metta. "Consumer Protection or Overprotection?" *Human Ecology* 28, No. 4 (Fall 2000): 5–8.

Wurzel, Rüdiger K. W. "Flying into Unexpected Turbulence: The German EU Presidency in the Environmental Field," *German Politics* 9, No. 3 (September 2000): 23–43.

Zaboli, Roberto, Giancarlo Barbiroli, Ricardo Leoncini, Massimilano Mazzanti, and Sandro Montressor. "Regulation and Innovation in the Area of End-of-Life Vehicles," European Commission JRC-IPTS and Enterprise DG Study of the Impact Of EU Regulation on Innovation Of European Industry, IDSE-CNR, EUR 19598 EN, Milan, Italy (March 2000). Accessed at http://ftp.jrc.es/EURdoc/eur19598en.pdf on August 16, 2010.

# Index

9780230109360